Chemical Laboratory Practice

Editors

F. L. Boschke W. Fresenius J. F. K. Huber E. Pungor
G. A. Rechnitz W. Simon Th. S. West

Pretsch Clerc Seibl Simon

Tables of
Spectral Data for
Structure Determination of
Organic Compounds

Translated from the German by
K. Biemann

Springer-Verlag
Berlin Heidelberg New York Tokyo 1983

P. D. Dr. Ernö Pretsch, Professor Dr. Joseph Seibl,
Professor Dr. Wilhelm Simon
Eidgenössische Technische Hochschule, Laboratorium für
Organische Chemie, Universitätsstraße 16, CH-8092 Zürich

Professor Dr. Thomas Clerc
Pharmazeutisches Institut der Universität, Sahlistraße 10,
CH-3000 Bern

Translator

Prof. Dr. Klaus Biemann
Dept. of Chemistry, Massachusetts Inst. of Technology,
Cambridge, MA 02139 / USA

Editors

Dr. Friedrich L. Boschke, Springer-Verlag, Postfach 105 280, D-6900 Heidelberg 1, FRG

Prof. Dr. Wilhelm Fresenius, Institut Fresenius, Chemische und Biologische
Laboratorien GmbH, Im Maisel, D-6204 Taunusstein 4, FRG

Prof. Dr. J. F. K. Huber, Institut für Analytische Chemie der Universität Wien,
Währinger Straße 38, A-1090 Wien, Austria

Prof. Dr. Ernö Pungor, Institute for General and Analytical Chemistry, Gellért-tér 4,
H-1502 Budapest XI, Hungary

Prof. Garry A. Rechnitz, Dept. of Chemistry, Univ. of Delaware, Newark, DE 19711,
USA

Prof. Dr. Wilhelm Simon, Eidgenössische Technische Hochschule, Laboratorium für
Organische Chemie, Universitätsstraße 16, CH-8092 Zürich, Switzerland

Prof. Thomas S. West, Macaulay Institute for Soil Research, Craigiebuckler,
Aberdeen AB9 2QJ, U.K.

English Translation of the revised 2nd German Edition 1981:
Anleitungen für die chemische Laboratoriumspraxis, Bd. XV
ISBN 3-540-10556-6 Springer-Verlag Berlin Heidelberg New York

ISBN 3-540-12406-3 Springer-Verlag Berlin Heidelberg New York Tokyo
ISBN 0-387-12406-3 Springer-Verlag New York Heidelberg Berlin Tokyo

Library of Congress Cataloging in Publication Data. Tabellen zur Strukturaufklärung
organischer Verbindungen mit spektroskopischen Methoden. English. Tables for structure determination of organic compounds by spectroscopic techniques. (Chemical
laboratory practice) "English translation of the revised 2nd German edition, 1981"–
Copyright p. Includes index. 1. Organic compounds–Spectra–Tables. 2. Chemical structure–Tables. I. Pretsch, Ernö, 1942- . III. Series. QC462.85.T313 1983. 547.1'22. 83-6819
ISBN 0-387-12406-3 (U.S.)

Printing and bookbinding: Beltz Offsetdruck, Hemsbach
2152/3140-543210

Preface to the English Edition

Although numerical data are, in principle, universal, the compilations
presented in this book are extensively annotated and interleaved with
text. This translation of the second German edition has been prepared
to facilitate the use of this work, with all its valuable detail, by
the large community of English-speaking scientists. Translation has
also provided an opportunity to correct and revise the text, and to
update the nomenclature. Fortunately, spectroscopic data and their
relationship with structure do not change much with time so one can
predict that this book will, for a long period of time, continue to
be very useful to organic chemists involved in the identification of
organic compounds or the elucidation of their structure.

Klaus Biemann
Cambridge, MA, April 1983

Preface to the First German Edition

Making use of the information provided by various spectroscopic techniques has become a matter of routine for the analytically oriented
organic chemist. Those who have graduated recently received extensive
training in these techniques as part of the curriculum while their
older colleagues learned to use these methods by necessity. One can,
therefore, assume that chemists are well versed in the proper choice
of the methods suitable for the solution of a particular problem and
to translate the experimental data into structural information.

Those who are not specialists in any of these techniques and therefore not in continuous contact with the corresponding data may wish
to have a compact summary of reference data in a form that can be
grasped easily. Even experts appreciate the opportunity to look up

in a summary information about compound types with which they are
not familiar. The tables compiled in this book are meant to fill this
gap. They were compiled for courses and exercises which the authors
offered over a ten year period to students at the Federal Institute
of Technology (ETH), Zurich, and are thus well suited as a basis for
similar courses elsewhere.

Considering such a broad effort there will undoubtedly be some omissions and errors in our presentations. We would be grateful to users
of this book for suggestions and criticisms which would help us to
keep it up to date. A postcard included in the book may make it easier
for the reader to make such comments. We would also be grateful to
receive reprints of papers containing information and data which could
be incorporated in later editions and thus improve their usefulness.

A book such as this could not be assembled without the help of enthusiastic and knowledgeable collaborators who contributed a good deal
to the work. Our special thanks go to Miss I. Port, Dr. D. Wegmann
as well as Mr. P. Oggenfuss and Dr. R. Schwarzenbach.

Preface to the Second German Edition

This second edition provided an opportunity to include many additions
and correct some errors. Amongst other improvements tables and figures
concerning opaque regions and error-signals in the infrared were added.
We also appreciate correspondence which led to a number of corrections.

Our special thanks go to Dr. D. Wegmann and Mr. P. Oggenfuss for their
very careful cooperation. It is due to their efforts that this new
edition was produced in a timely fashion.

Table of Contents

Proton Resonance Spectroscopy H5

Infrared Spectroscopy I5

Introduction

The following collection of data is intended to serve as an aid in
the interpretation of ^{13}C- and ^{1}H-nuclear magnetic resonance, infrared, mass and electron excitation spectra. It is to be viewed as an
addition to texts and reference works dealing with these spectroscopic
techniques. It is designed for those who are routinely faced with the
task to interprete this type of spectral information. The use of this
book for the interpretation of spectra requires only the knowledge of
basic principles of these techniques, but its content is structured
in a way that it will also serve as a reference work for the special-
ist.

To aid rapid access to relevant data the following codes are listed
at the top of the appropriate pages:

KOMB: Summary of characteristic spectroscopic data arranged by structural elements.
   (pages B5 through B265)

^{13}C-NMR: ^{13}C-nuclear magnetic resonance spectral data ordered by
   compound types.
   (pages C5 through C265)

^{1}H-NMR: Proton magnetic resonance spectral data.
   (pages H5 through H370)

IR: Infrared absorption frequencies ordered by functional groups or
   compound types.
   (pages I5 through I280)

MS: Tables and suggestions for the interpretation of mass spectra.
   (pages M5 through M170)

UV/VIS: Tables, suggestions and reference spectra for rationalization
   of electron excitation spectra.
   (pages U5 through U155)

The tables are arranged as much as possible in an analogous manner
for all methods, but the details of the presentation are dictated by
the characteristics of the individual technique.

The pages are numbered in increments of five which should facilitate
later additions.

The summaries presented on pages B5 through B70 facilitate the recog-
nition of first-cut conclusions concerning the structure by those
users less familiar with such interpretations, particularly in those
cases where no ancillary information is available. The tables presented on pages B75 through B245 make it possible to check in a second
step the suspected structural elements for the most important compound
types. The remaining tables make it possible to predict the spectroscopic characteristics of a proposed compound. They also serve as a
reference compilation for the correlation of the structure of organic
compounds with the corresponding spectral data.

Because a large part of the tabulated data is from our own measurements and the rest is based on a large body of literature data a com-
prehensive reference to published sources is not included. Whenever
possible the data refer to conventional modes and conditions of meas-
urement. For example, the chemical shifts for NMR spectra were determined generally in deuterochloroform or carbon tetrachloride. The
wave numbers (IR) refer to solvents of low polarity, such as chloroform or carbon disulfide. Mass spectral data were recorded at an
electron energy of 70 eV. Most of the data were taken from the following sources:

Batterham, T.J.: NMR spectra of simple heterocycles. New York – London –
   Sydney – Toronto: Wiley-Interscience 1973.
Bhacca, N.S., Hollis, D.P., Johnson, L.F., Pier, E.A., Shoolery, J.N.:
   NMR spectra catalog. Varian Associates 1962 and 1963.
Bremser, W., Ernst, L., Franke, B.: Carbon-13 NMR spectral data. 2nd
   ed. Weinheim: Verlag Chemie 1979.
Bruegel, W.: Handbook of NMR spectral parameters, Vol. 1 – 3. London –
   Philadelphia – Rheine: Heyden 1979.
Clerc, J.T., Pretsch, E.: Kernresonanzspektroskopie, Teil I: Protonen-
   resonanz. 2nd ed. Frankfurt/Main: Akademische Verlagsgesellschaft
   1973.

Clerc, J.T., Pretsch, E., Sternhell, S.: ^{13}C-Kernresonanzspektroskopie. Frankfurt/Main: Akademische Verlagsgesellschaft 1973.

Colthup, N.B., Daly, L.H., Wiberley, S.E.: Introduction to infrared and raman spectroscopy. New York - London: Academic Press 1964.

Cross, A.D.: Introduction to practical infrared spectroscopy. London: Butterworths 1960.

Dokumentation der Molekülspektroskopie. Institut für Spektrochemie und angewandte Spektroskopie, Dortmund. Weinheim: Verlag Chemie.

Dolphin, D., Wick, A.E.: Tabulation of infrared spectral data. New York - London - Sydney - Toronto: John Wiley & Sons 1977.

Graselli, J.G., Ritchey, W.M. (Eds.): Atlas of spectral data and physical constants for organic compounds. Cleveland: CRC Press 1975.

Hediger, H.J.: Infrarotspektroskopie. Frankfurt/Main: Akademische Verlagsgesellschaft 1971.

Jackman, L.M., Sternhell, S.: Applications of NMR spectroscopy in organic chemistry. 2nd ed. Oxford - London - Edinburgh - New York - Toronto - Sydney - Paris - Braunschweig: Pergamon Press, 2nd Ed., 1969.

Johnson, L.F., Jankowski, W.C.: Carbon-13 NMR spectra. Huntington - New York: R.E. Krieger Publ. Co. 1978.

Kirba-Kartei. Eglisau: Verlag Dr. H.J. Hediger.

NMR Spectra Catalog. Philadelphia: Sadtler Research Laboratories.

Seibl, J.: Massenspektrometrie. Frankfurt/Main: Akademische Verlagsgesellschaft 1970.

Socrates, G.: Infrared characteristic group frequencies. Chichester - New York - Brisbane - Toronto: John Wiley & Sons 1980.

Yukawa, Y.: Handbook of organic structural analysis. New York - Amsterdam: W.A. Benjamin 1965

Abbreviations and Symbols

al	aliphatic
ar	aromatic
as	asymmetric
ax	axial
comb	combination frequency
δ	IR: deformation frequency NMR: chemical shift
eq	equatorial
γ	skeletal vibration
gem	geminal
hal	halogen
ip	"in plane" vibration
oop	"out of plane" vibration
st	stretching vibration
sy	symmetric

Summary of the Regions of the Chemical Shifts for Carbon in Various Bonding Environments
(δ in ppm relative to TMS)

The multiplicity of "off-resonance" uncoupled first order spectra is indicated by the
following abbreviations: S = singlet, D = doublet, T = triplet, Q = quadruplet

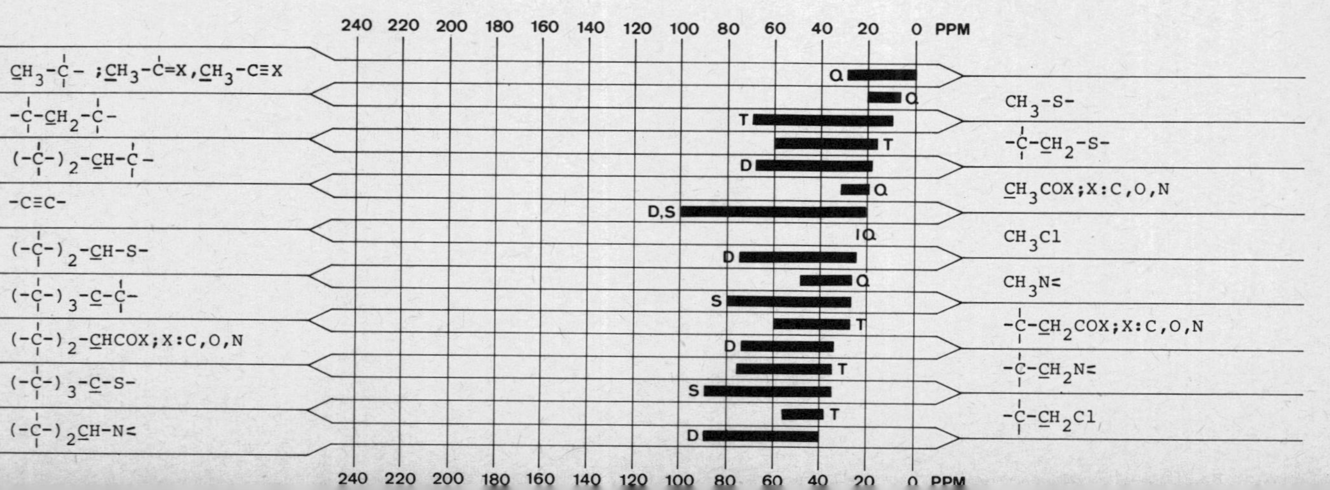

^{13}C-NMR, SUMMARY

PPM scale: 0 20 40 60 80 100 120 140 160 180 200 220 240

Upper chart labels (top):

- $(-\overset{|}{\underset{|}{C}}-)_3-\underline{C}-COX; X: C,O,N$
- $(-\overset{|}{\underset{|}{C}}-)_3-\underline{C}-N<$
- CH_3-NO_2
- $(-\overset{|}{\underset{|}{C}}-)_2-\underline{C}H-O-$
- $(-\overset{|}{\underset{|}{C}}-)_3-\underline{C}-Cl$
- $(-\overset{|}{\underset{|}{C}}-)_3-\underline{C}-O-$
- $\underline{C}H_2=C<$
- $>C=C<$
- ring $\underline{C}-X$ X: any substituent
- $-C\equiv N$
- $>\underline{C}=N-X; X: C,O$
- $\alpha,\beta-unsat. \underline{C}OOH$
- $-\overset{|}{\underset{|}{C}}-\underline{C}OOH$
- $-\overset{|}{\underset{|}{C}}-\underline{C}S-X; X: O,N$
- $-\overset{|}{\underset{|}{C}}-\underline{C}HO$
- $-\overset{|}{\underset{|}{C}}-\underline{C}S-\overset{|}{\underset{|}{C}}-$

Lower chart labels (bottom):

- CH_3-O-
- $(-\overset{|}{\underset{|}{C}}-)_2-\underline{C}H-Cl$
- $-\overset{|}{\underset{|}{C}}-\underline{C}H_2-O-$
- $-\overset{|}{\underset{|}{C}}-\underline{C}H_2-NO_2$
- $(-\overset{|}{\underset{|}{C}}-)_2-\underline{C}H-NO_2$
- $(-\overset{|}{\underset{|}{C}}-)_3-\underline{C}-NO_2$
- ring $\underline{C}-H$ X: any substituent
- $-O-\overset{|}{\underset{|}{C}}-O-$
- pyridine-type   X: any substituent
- pyridine-type   X: any substituent
- $\alpha,\beta-unsat. \underline{C}OX; X: O,N,Cl$
- $-\overset{|}{\underset{|}{C}}-\underline{C}OX; X: O,N,Cl$
- $\alpha,\beta-unsat. \underline{C}OH$
- $\alpha,\beta-unsat. >\underline{C}O$
- $-\overset{|}{\underset{|}{C}}-\underline{C}O-\overset{|}{\underset{|}{C}}-$

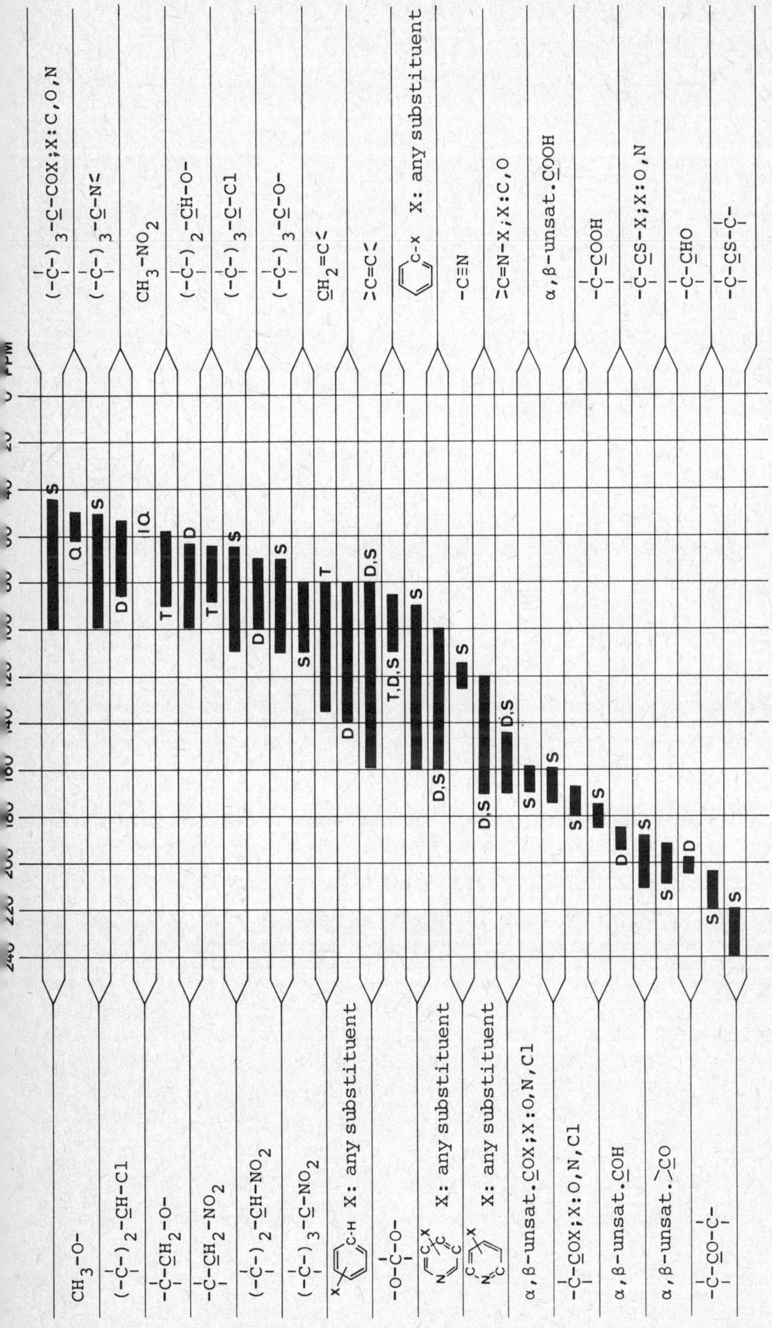

KOMB

^{1}H-NMR, SUMMARY

<u>Summary of the Regions of the ^{1}H Chemical Shifts of Protons in Various Bonding Environments</u> (δ in ppm relative to TMS)

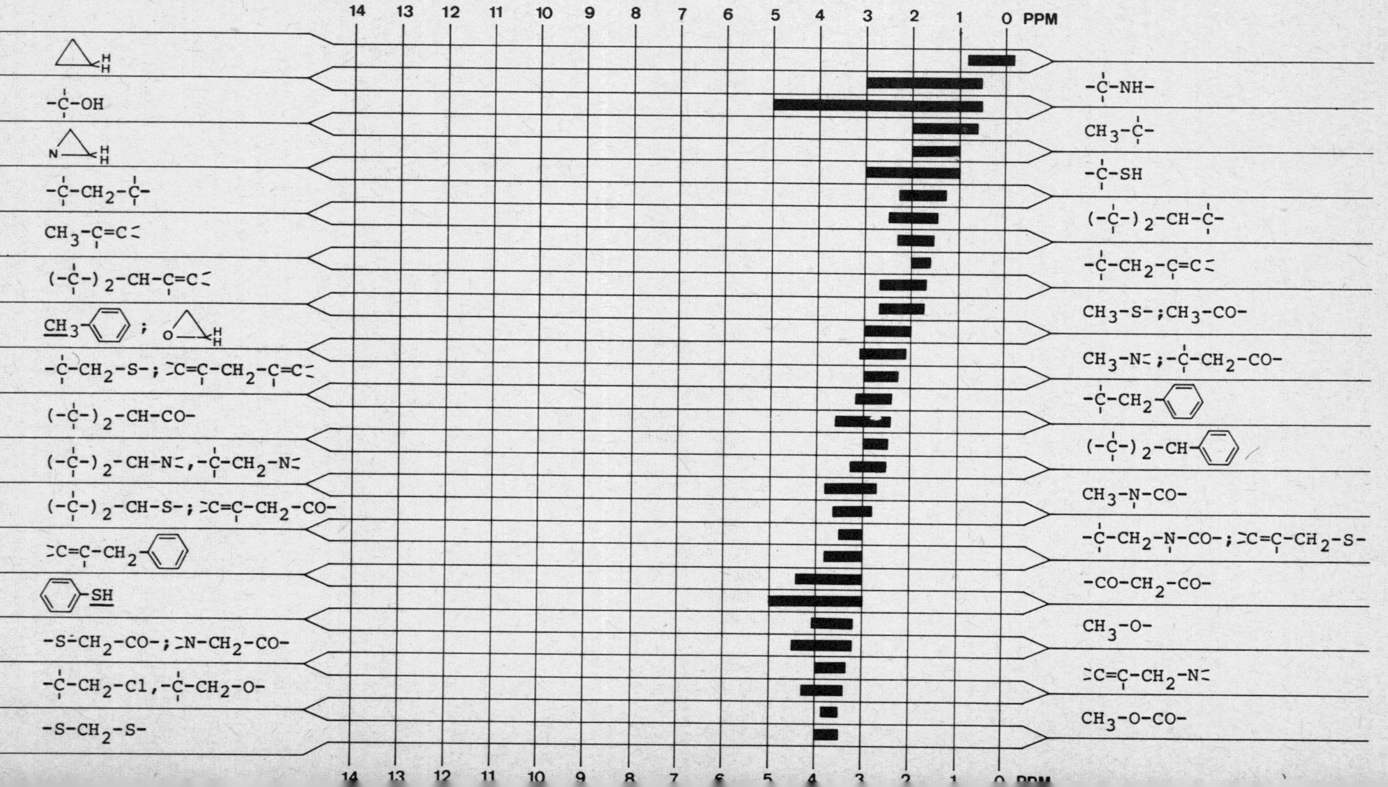

B15

^{1}H-NMR, SUMMARY

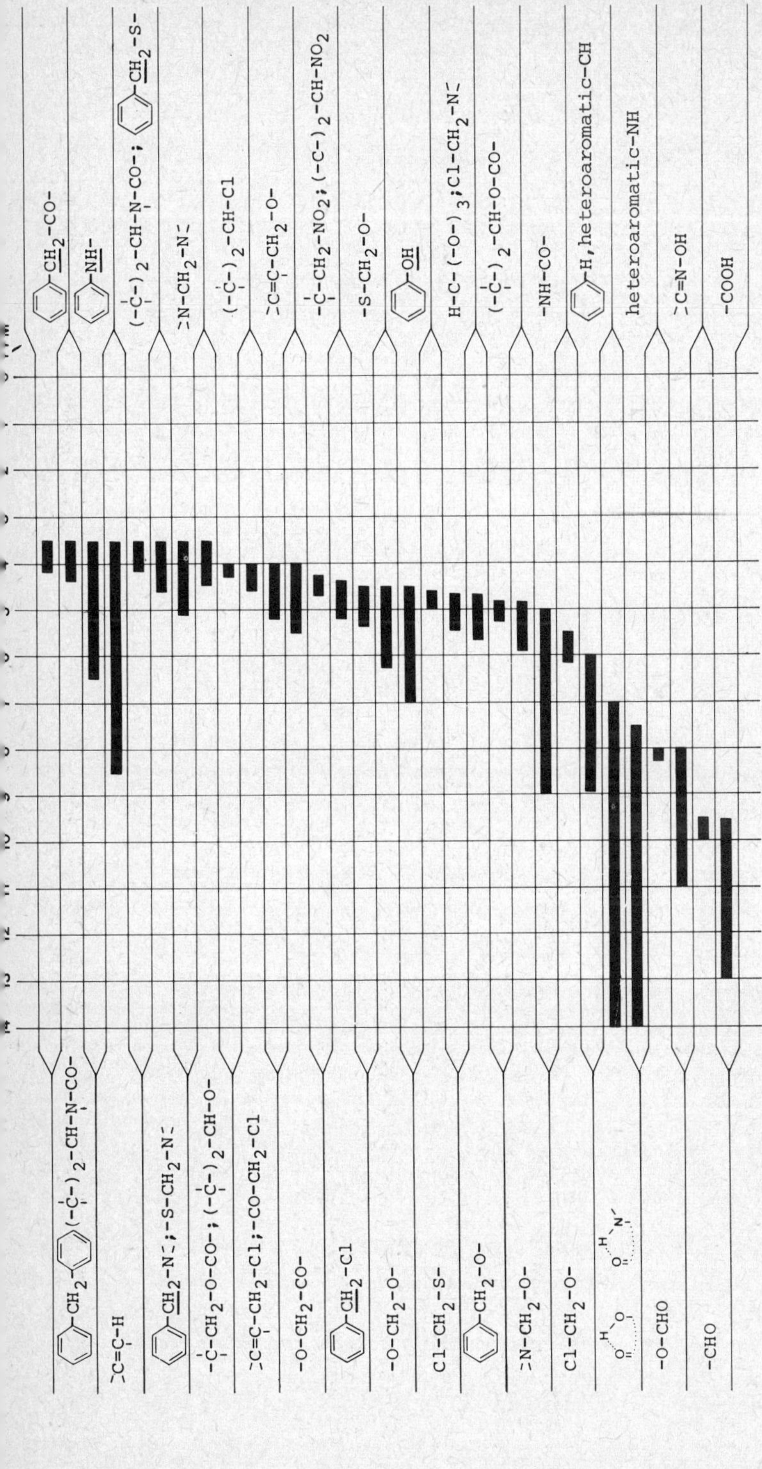

$\underline{CH_2}$-CO- (phenyl)
$\underline{NH}$- (phenyl)
(-C=)$_2$-CH-N-CO- ; $\underline{CH_2}$-S- (phenyl)
$\geq$N-CH$_2$-N$\leq$
(-C-)$_2$-CH-Cl
$\geq$C=C-CH$_2$-O-
-C-CH$_2$-NO$_2$; (-C-)$_2$-CH-NO$_2$
-S-CH$_2$-O-
$\underline{OH}$ (phenyl)
H-C-(-O-)$_3$; Cl-CH$_2$-N$\leq$
(-C-)$_2$-CH-O-CO-
-NH-CO-
H, heteroaromatic-CH (phenyl)
heteroaromatic-NH
$\geq$C=N-OH
-COOH

O 1 2 3 4 5 6 7 8 9 10 11 12 13 14 PPM

CH$_2$ (phenyl) (-C-)$_2$-CH-N-CO-
$\geq$C=C-H
$\underline{CH_2}$-N$\leq$; -S-CH$_2$-N$\leq$
-C-CH$_2$-O-CO- ; (-C-)$_2$-CH-O-
$\geq$C=C-CH$_2$-Cl ; -CO-CH$_2$-Cl
-O-CH$_2$-CO-
$\underline{CH_2}$-Cl (phenyl)
-O-CH$_2$-O-
Cl-CH$_2$-S-
$\underline{CH_2}$-O- (phenyl)
$\geq$N-CH$_2$-O-
Cl-CH$_2$-O-
-O-CHO
-CHO

KOMB

^{1}H-NMR, SUMMARY

<u>Summary of Ranges of Coupling Constants Between Differently Bound Protons</u> (|J| in Hz)

X and Y represent different substituents.

<u>Geminal coupling:</u>

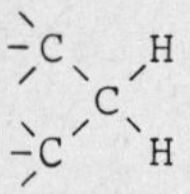

10 - 14 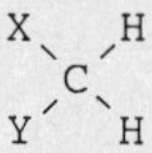   0 - 20 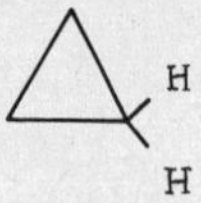   2 - 5 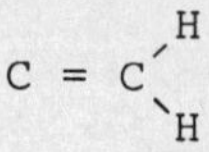   0 - 4

<u>Vicinal coupling:</u>

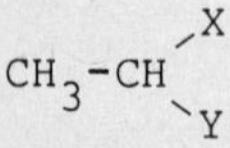

CH_3-CH X/Y $C - C$ ax,ax: 6 - 13
6 - 8   0 - 18   ax,eq: 2 - 4
                                        eq,eq: 2 - 4

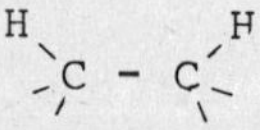

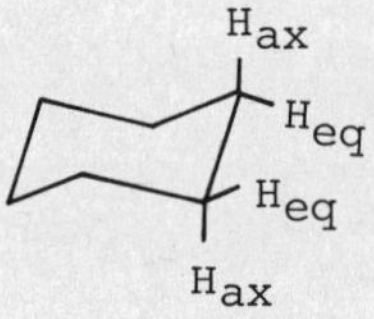

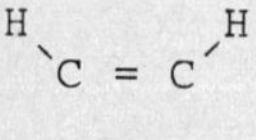

5 - 14 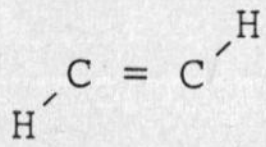   12 - 18 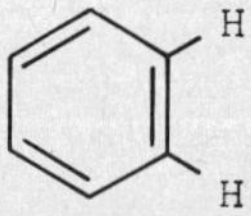   6 - 10

<u>"Long-range" coupling:</u>

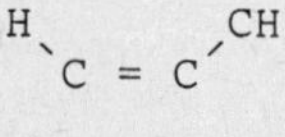

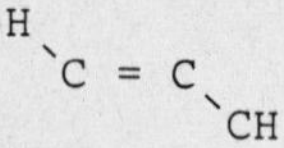

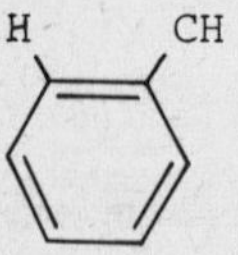

0 - 2.5

$CH - C \equiv CH$
$CH - C \equiv C - CH$ 2 - 3

Summary of the Most Important Strong IR Absorption Bands (in cm^{-1})

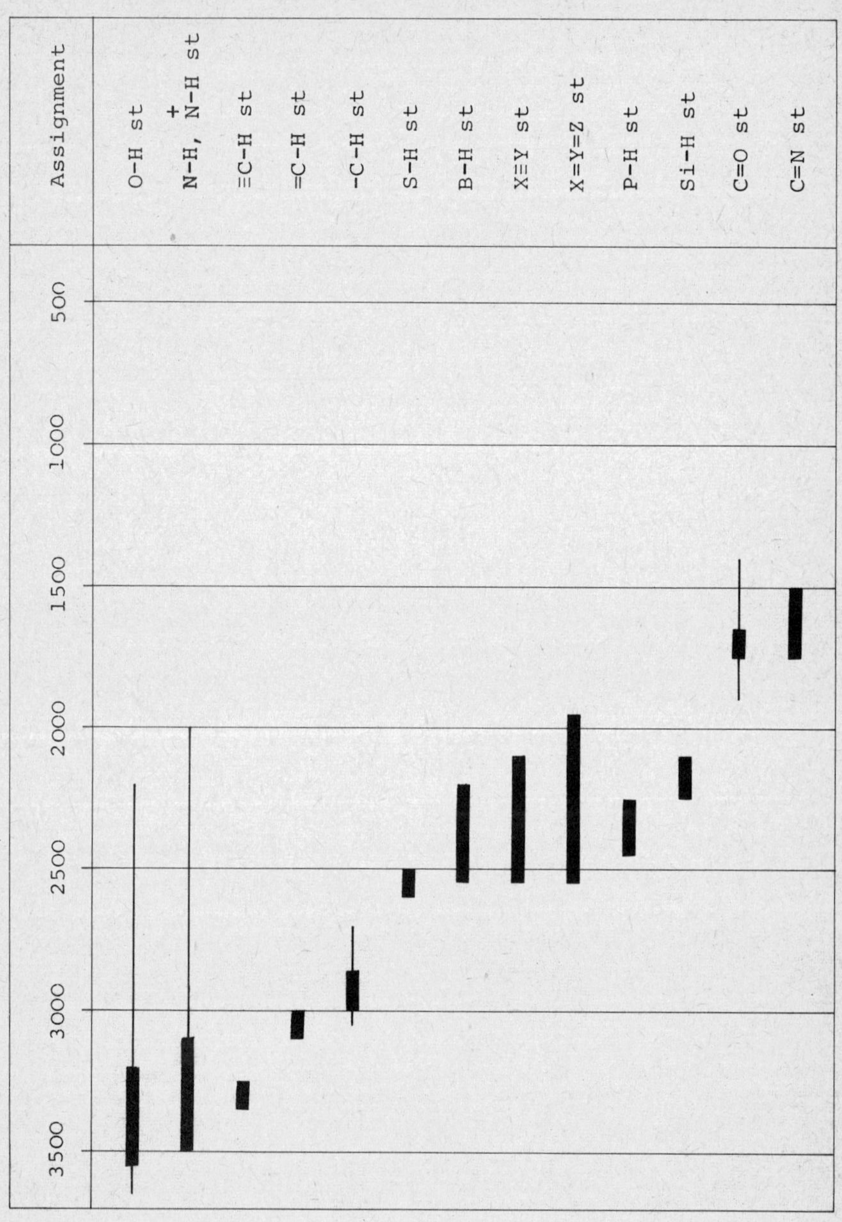

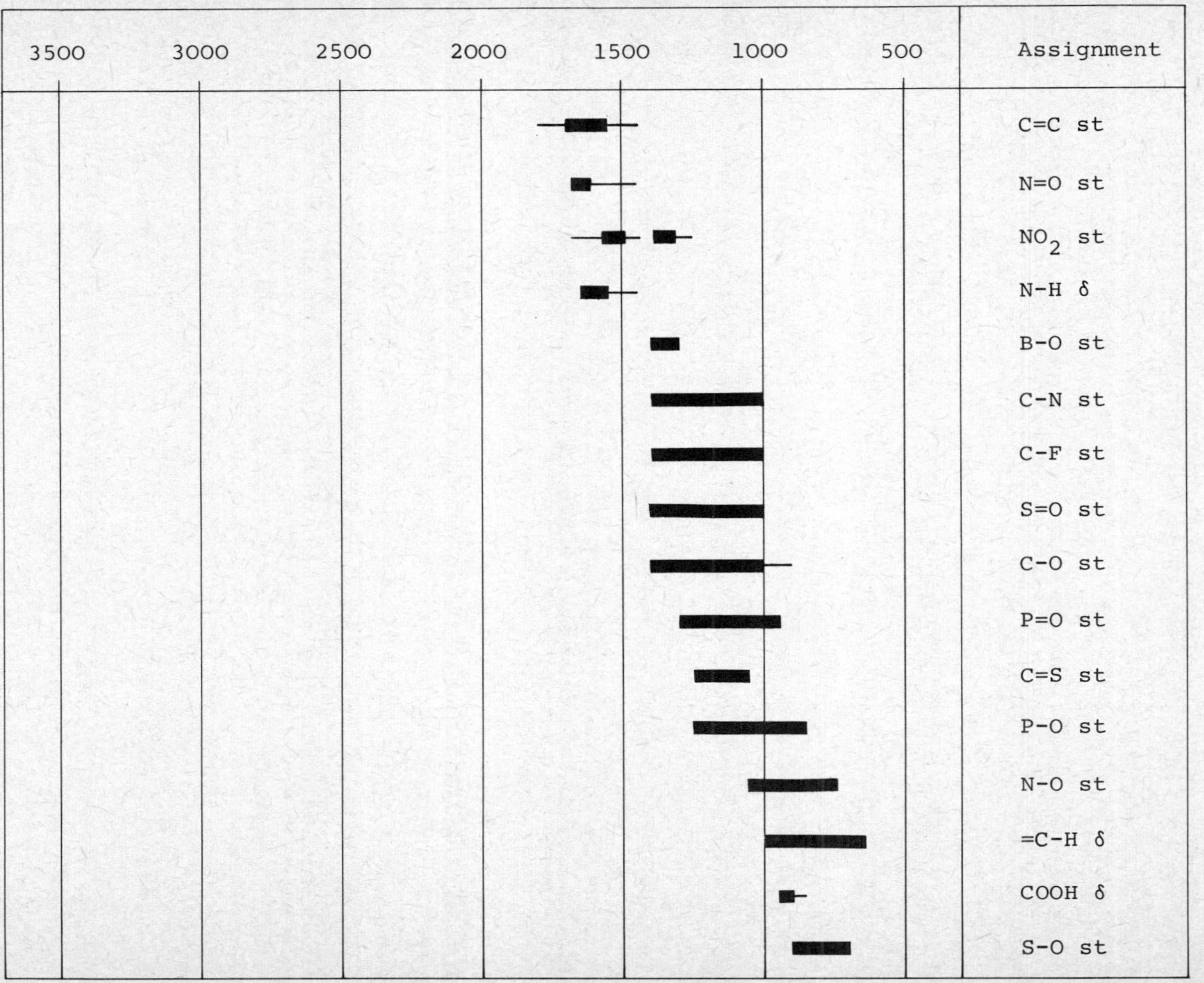

3500
3000
2500
2000
1500
1000
500
Assignment
C=C st
N=O st
NO₂ st
N–H δ
B–O st
C–N st
C–F st
S=O st
C–O st
P=O st
C=S st
P–O st
N–O st
=C–H δ
COOH δ
S–O st

Summary of IR Absorption Bands of $>$C=O Groups (in cm^{-1})

X: any substituent

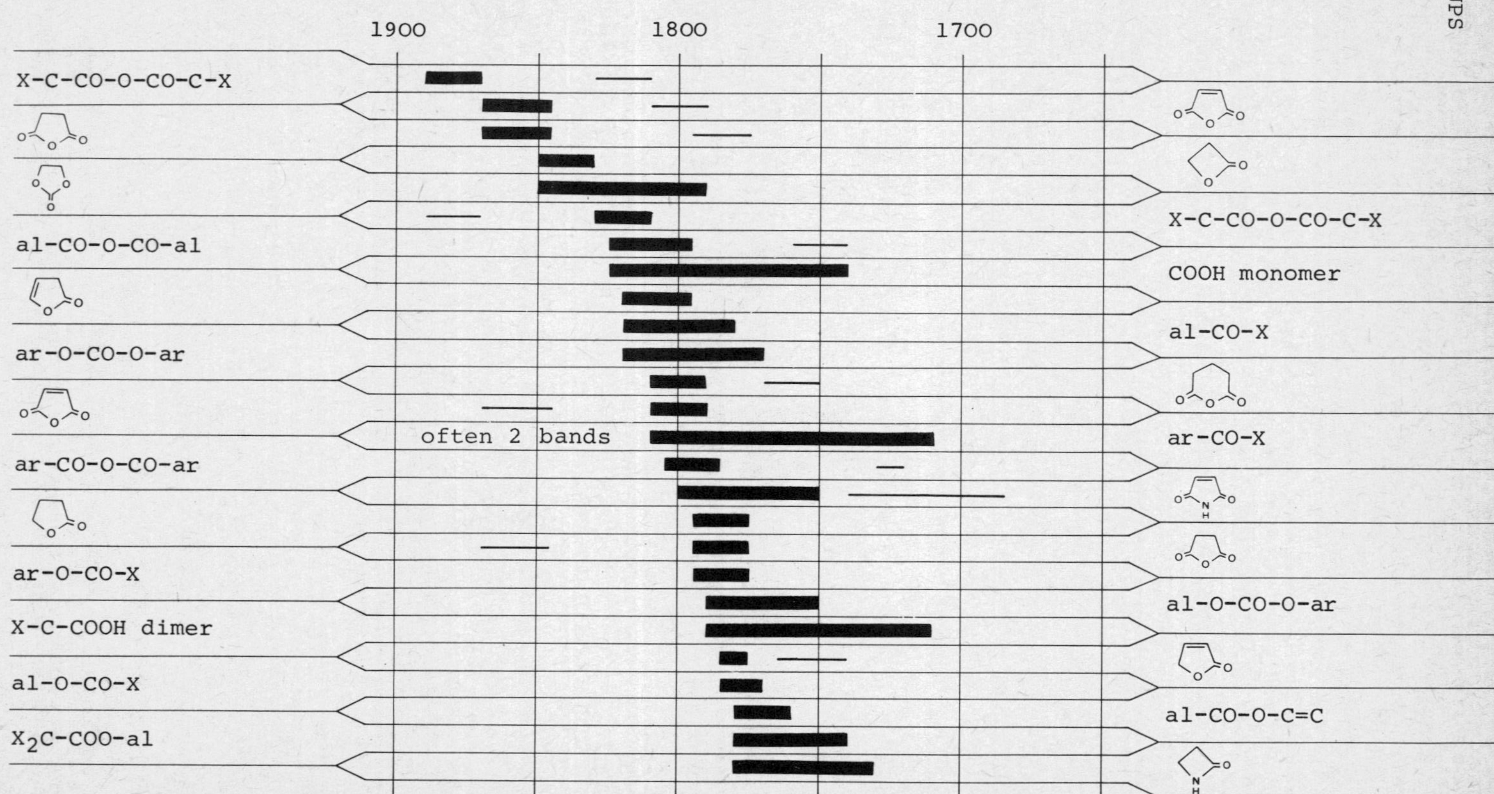

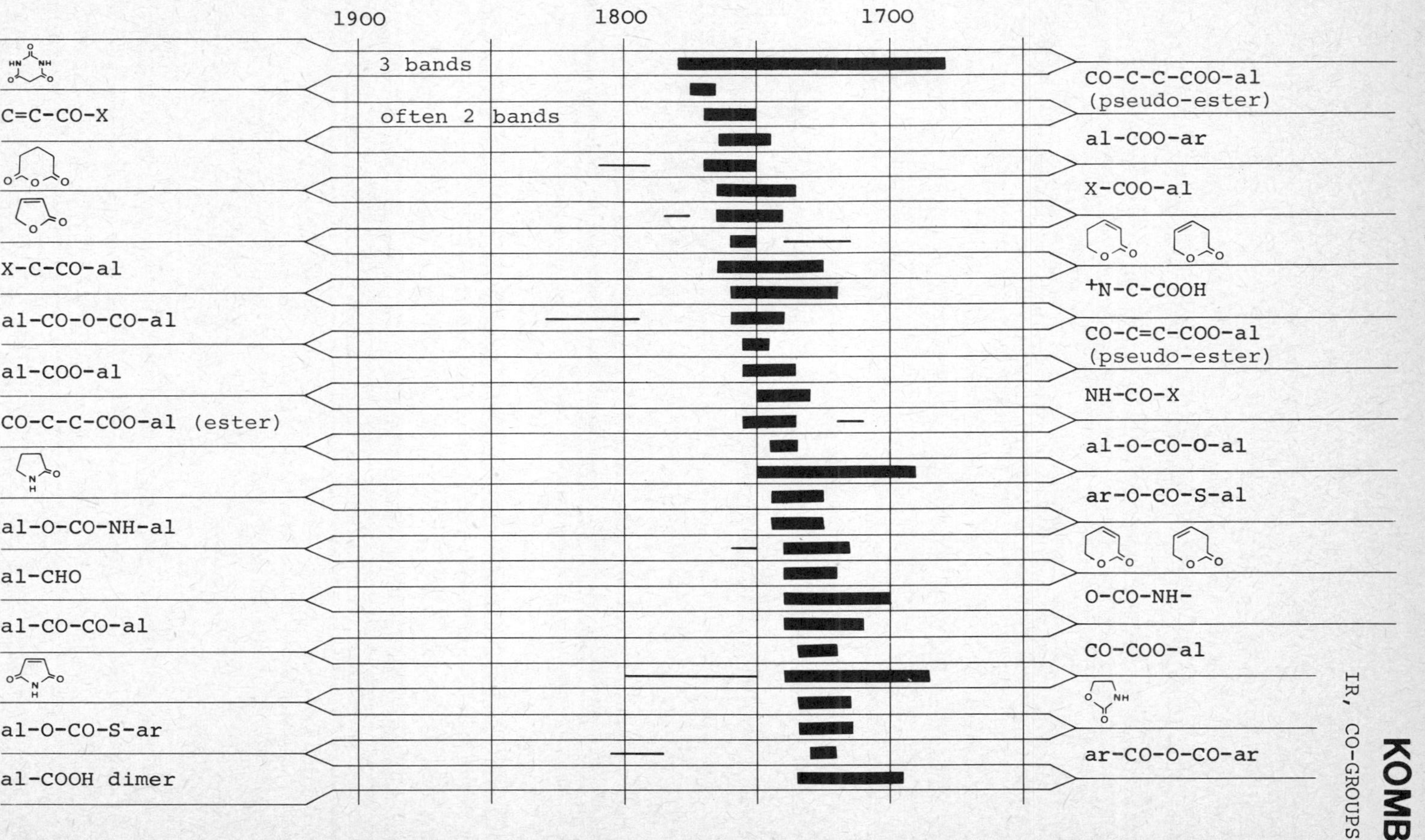
1900
1800
1700
3 bands
often 2 bands
C=C-CO-X
X-C-CO-al
al-CO-O-CO-al
al-COO-al
CO-C-C-COO-al (ester)
al-O-CO-NH-al
al-CHO
al-CO-CO-al
al-O-CO-S-ar
al-COOH dimer
CO-C-C-COO-al (pseudo-ester)
al-COO-ar
X-COO-al
+N-C-COOH
CO-C=C-COO-al (pseudo-ester)
NH-CO-X
al-O-CO-O-al
ar-O-CO-S-al
O-CO-NH-
CO-COO-al
ar-CO-O-CO-ar

KOMB

IR, CO-GROUPS

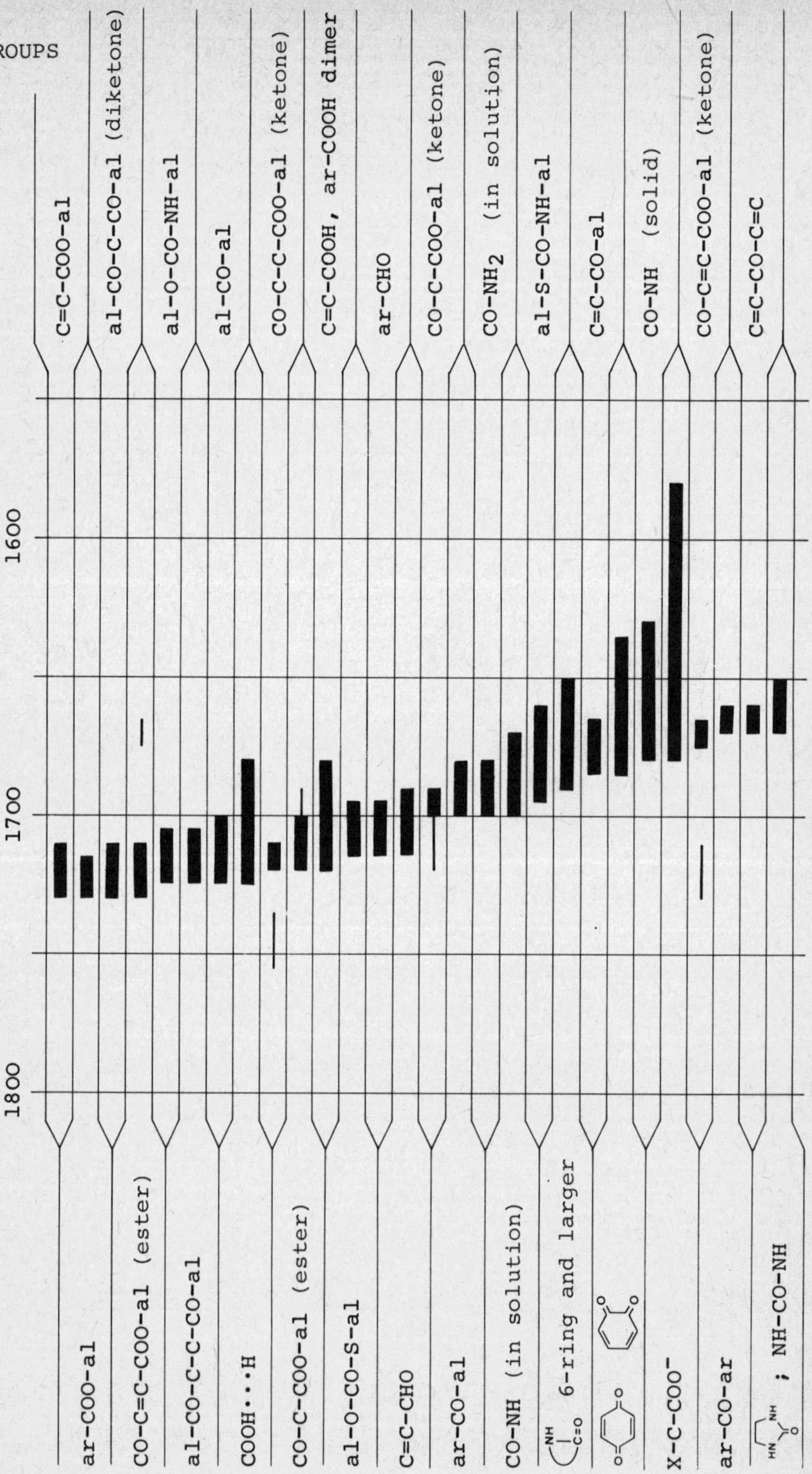

1600
1700
1800
C=C-COO-al
al-CO-C-CO-al (diketone)
al-O-CO-NH-al
al-CO-al
CO-C-C-COO-al (ketone)
C=C-COOH, ar-COOH dimer
ar-CHO
CO-C-COO-al (ketone)
CO-NH2 (in solution)
al-S-CO-NH-al
C=C-CO-al
CO-NH (solid)
CO-C=C-COO-al (ketone)
C=C-CO-C=C
ar-COO-al
CO-C=C-COO-al (ester)
al-CO-C-C-CO-al
COOH···H
CO-C-COO-al (ester)
al-O-CO-S-al
C=C-CHO
ar-CO-al
CO-NH (in solution)
6-ring and larger
X-C-COO⁻
ar-CO-ar
; NH-CO-NH

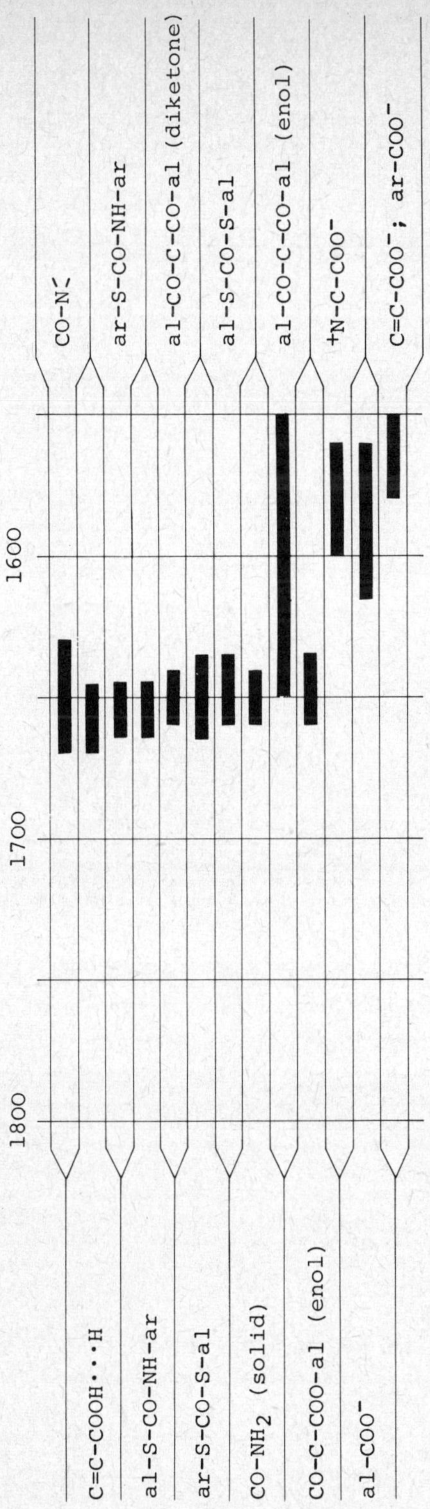
CO-N<
ar-S-CO-NH-ar
al-CO-C-CO-al (diketone)
al-S-CO-S-al
al-CO-C-CO-al (enol)
+N-C-COO⁻
C=C-COO⁻ ; ar-COO⁻
C=C-COOH···H
al-S-CO-NH-ar
ar-S-CO-S-al
CO-NH₂ (solid)
CO-C-COO-al (enol)
al-COO⁻
1600
1700
1800

UV/VIS-Absorption Bands of Various Compound Types

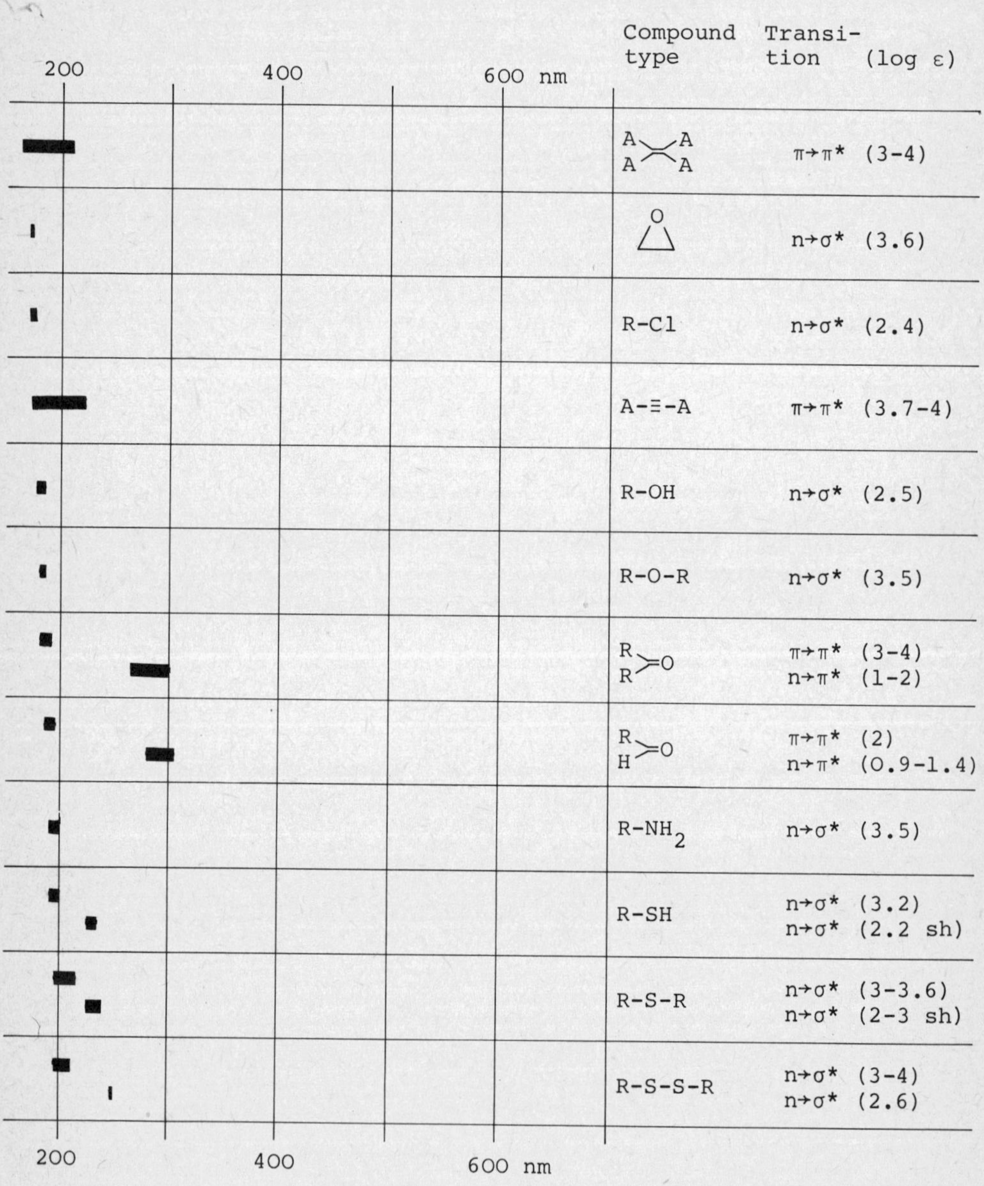

A: alkyl or H R: alkyl sh: shoulder

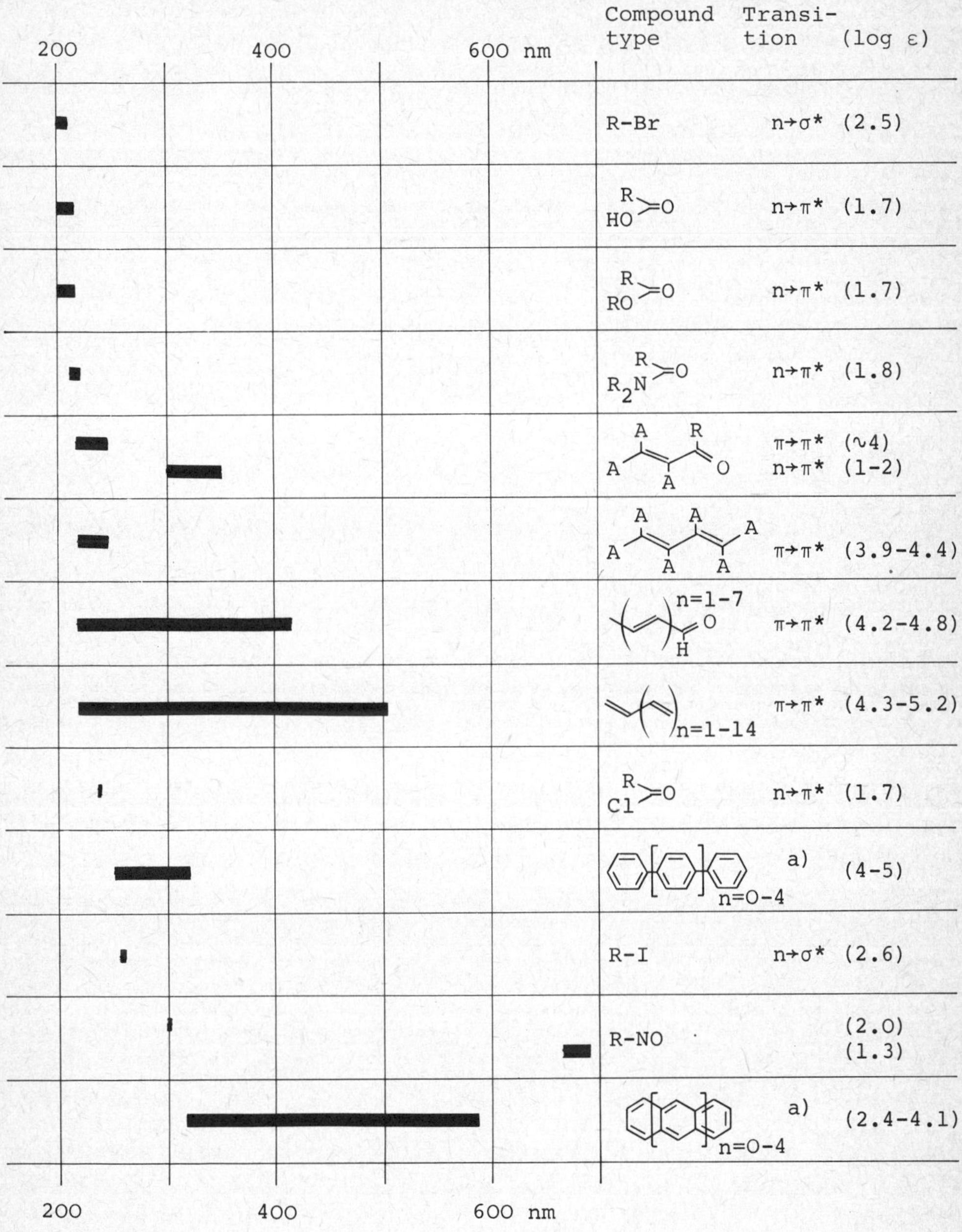

	Compound type	Transition	(log ε)
	R–Br	n→σ*	(2.5)
	R, HO—C=O	n→π*	(1.7)
	R, RO—C=O	n→π*	(1.7)
	R, R₂N—C=O	n→π*	(1.8)
	A, R, A, A—C=O (conjugated)	π→π* (∼4); n→π*	(1-2)
	A, A, A, A, A (diene)	π→π*	(3.9-4.4)
	polyene aldehyde, n=1-7	π→π*	(4.2-4.8)
	polyene, n=1-14	π→π*	(4.3-5.2)
	R, Cl—C=O	n→π*	(1.7)
	polyphenyl, n=0-4	a)	(4-5)
	R–I	n→σ*	(2.6)
	R–NO		(2.0); (1.3)
	polyacene, n=0-4	a)	(2.4-4.1)

Horizontal axis (top and bottom): 200, 400, 600 nm

a) longest wavelength absorption maximum

KOMB

ALKANES, CYCLOALKANES

Characteristic Spectroscopic Data for Alkanes and Saturated Alicyclics

	Assignment	Range	Comments	Details see page:
^{13}C-NMR			reference data: alkanes	C5
			cycloalkanes	C50
	$-CH_3$		$-CH_3$, $-CH_2-$, $>CH-$ and $>C<$ can be differentiated	
	$-CH_2-$	5-60 ppm	by partial ("off resonance") uncoupled spectra	
	$>CH-$		or based on relaxation times	
	$>C<$		found outside of the quoted range if present in three-membered rings	
^{1}H-NMR			reference data, alkanes: chemical shifts	H5
			coupling constants	H20
			cycloalkanes: chemical shifts and coupling constants	H185
	$-CH_3$	0.8-1.2 ppm		
	$-CH_2-$	1.1-1.8 ppm	found outside of the quoted range if present in three-membered rings	
	$>CH-$			

IR		reference data	I5
$\gtrless$CH st	3000–2840 cm^{-1}	found outside the quoted range if present in three-membered rings	
-CH$_3$ δ as -CH$_2$- δ	∿1460 cm^{-1}		
-CH$_3$ δ sy	∿1380 cm^{-1}	doublet for geminal methyl groups	
-CH$_2$- γ	770–720 cm^{-1}	in C-(CH$_2$)$_n$-C if n ⩾ 4 at ∿720 cm^{-1}	
MS Molecular ion		n-alkanes: weak isoalkanes: very weak $\Big\}$ m/z = 14n + 2 monocyclo-alkanes: medium m/z = 14n	M115
Fragments		n-alkanes: local maxima at 14n + 1, intensity variations smooth, minimum at M† – 15 isoalkanes: local maxima at 14n + 1, intensity distribution irregular, relative maxima due to fragmentation at branching points with charge retention at most substituted C. monocyclo-alkanes: local maxima at 14n – 1, intensity distribution irregular, relative maxima due to cleavage at ring.	
Rearrange-ments		n-alkanes: non-specific isoalkanes monocyclo-alkanes $\Big\}$ elimination of alkanes $\Big\{$ m/z = 14n m/z = 14n – 2	
UV		no absorption above 200 nm	

KOMB

ALKENES, CYCLOALKENES

Characteristic Spectroscopic Data for Alkenes and Saturated Alicyclics

	Assignment	Range	Comments	Details see page:
^{13}C-NMR	C=C	100-150 ppm	reference data	C80, C100
	C-(C=C)	10-40 ppm	considerable differences between $C{-}C{=}C{<}^{X}_{H}$ and $C{-}C{=}C{<}^{H}_{X}$	C10
^{1}H-NMR	H-(C=C)	4.5-6 ppm	reference data, additivity rules coupling constants |J|: geminal 0- 3 Hz cis 5-12 Hz trans 12-18 Hz	H205
	CH$_3$-(C=C)	∼1.7 ppm	reference data	H5, H225
	-CH$_2$-(C=C)	∼2.0 ppm	alkenes	H5
			cycloalkenes	H230
			considerable differences between $-CH_2-C-C{=}C-$ (∼1.5 ppm) and $CH_2-C-C{=}C$ (with C_n) n⩾1 (∼1.8 ppm, see H230)	
			coupling constants |J|: $CH_2{-}CH{=}C$ ∼7 Hz $CH_2{-}CH{=}C$ (with C_n) n=2 ∼0.5 Hz; n=3 ∼1.5 Hz; n=4 ∼4 Hz	
			"long-range" coupling constants: see H205	

IR		reference data	I20
	H–C(=C) st	$3100–3000\ cm^{-1}$	
	C=C st	$1690–1635\ cm^{-1}$	
	H–C(=C) δ oop	$1000–675\ cm^{-1}$	
		$-CH_2-(C=C)$ δ at $1440\ cm^{-1}$ (see I10)	
MS	Molecular ion	medium abundant, m/z = 14n, in monocyclic alkenes at 14n − 2	M115
	Fragments	local maxima at 14n − 1, in monocyclic alkenes at 14n − 3	
		double bonds can be localized only in special cases	
	Rearrangements	unspecific; in cyclohexenes Retro-Diels-Alder-reaction:	
		specific for: following scheme	
UV	C=C π→π*	<210 nm (log ε = 3–4)	U30
	$(C=C)_2$ π→π*	215–280 nm (log ε=3.5–4.5)	
		for isolated double bonds; for highly substituted double bonds often endabsorption	
		reference data	

Characteristic Spectroscopic Data of Alkynes

	Assignment	Range	Comments	Details see page:
^{13}C-NMR	$C \equiv C$	65–85 ppm	reference data coupling constant $H-C\equiv^{13}C$: $\sim$50 Hz, generally recognizable in partially ("off resonance") uncoupled spectra	C110
	$C-(C\equiv C)$	0–30 ppm		C10
^{1}H-NMR	$H-C\equiv C$	1.5–3 ppm	reference data coupling constants $|J|$: $\quad$ CH–C$\equiv$CH $\quad\sim$3 Hz $\qquad\qquad\qquad\qquad\qquad$ CH–C$\equiv$C–CH $\quad\sim$3 Hz	H225
	$CH_3-C\equiv C$	$\sim$1.8 ppm		H225
	$-CH_2-C\equiv C$	$\sim$2.2 ppm		
	$\rangle CH-C\equiv C$	$\sim$2.6 ppm		
IR			reference data	I40
	$H-(C\equiv C)$ st	3340–3250 cm^{-1}	sharp	
	$C\equiv C$ st	2260–2100 cm^{-1}	sometimes very weak	

MS	Molecular ion		weak, for 1-alkynes up to C_7 often absent	M115
	Fragments Rearrange- ments }		vary in extent between alkanes and aromatics	
UV	$C \equiv C$ $\pi \to \pi^*$	<210 nm ($\log \varepsilon = 3.7-4.0$)	endabsorption, often a few weak bands <240 nm	

Characteristic Spectroscopic Data of Aromatic Hydrocarbons

	Assignment	Range	Comments	Details see page:
^{13}C-NMR			reference data	C115
	arC	120–150 ppm	also in polycyclic aromatic hydrocarbons	C120
	arCH	110–130 ppm		
	C–(arC)	10–60 ppm		C10
^{1}H-NMR			reference data	H245
	H–(arC)	6.8–7.5 ppm	in polycyclic aromatic hydrocarbons up to ∿9 ppm	
			coupling constants: J_{ortho} ∿7 Hz $\quad$ J_{meta} ∿2 Hz $\quad$ J_{para} <1 Hz (in routine-spectra generally not detectable)	H245
	CH$_3$–(arC)	∿2.3 ppm	often band broadening due to "long-range" coupling with aromatic protons (see H245)	H5,H30
	–CH$_2$–(arC)	∿2.6 ppm		H5,H30
	⟩CH–(arC)	∿2.9 ppm		H5
			geminal coupling in –CH$_2$–(arC): see H20	
IR			reference data	I45
	arC–H st	3080–3030 cm^{-1}	often multiple bands, weak	
	comb	2000–1650 cm^{-1}	very weak	
	arC–C st	∿1600 cm^{-1}	often split, sometimes not all three bands observable	
		∿1500 cm^{-1}		
		∿1450 cm^{-1}		
	arC–H δ oop	900–650 cm^{-1}	strong, frequently multiplicity of bands	

MS	Molecular ion		abundant, often base peak	
	Fragments		m/z = 39, 50-53, 63-65, 75-78; dehydrations; $M^{+\cdot} - 26$, $M^{+\cdot} - 39$; frequently doubly charged fragment ions; benzylic cleavage	M120
			m/z = 90-92	
			m/z = 127	
			m/z = 152, 153	
			m/z = 152, 165	
	Rearrangements			
UV			reference data	U50,U40, U110-U125
		$\sim$200-210 nm (log ε = $\sim$4) $\sim$260 nm (log ε = $\sim$2.4)	in benzene and alkyl benzene	

Characteristic Spectroscopic Data of Heteroaromatics

	Assignment	Range	Comments	Details see page:
^{13}C-NMR			reference data	C135–C165
	arC-(arX) } arC-(arC) }	100–160 ppm		
^{1}H-NMR			reference data	H265–H350
	H-(arC)	6–9 ppm	coupling in 6-membered rings similar to aromatic hydrocarbons (see H275); for 5-membered ring heteroaromatics see H265	
			H-(arN): 7–14 ppm, strongly solvent dependent, generally broad	
IR			reference data	I45
	arC-H st	3100–3000 cm^{-1}	frequently multiple bands, weak	
	arC-C st	∿1600 cm^{-1} } ∿1500 cm^{-1} } ∿1450 cm^{-1} }	often split, sometimes not all three bands observable	
	arC-H δ oop	1000–650 cm^{-1}	often strong, frequently multiplicity of bands arN-H st: 3500–2800 cm^{-1}	

MS		hints for heteroatoms	M125
Molecular ion		abundant, often base peak	
Fragments		m/z 39, 50-53 ... , [pyridine]+ : m/z 78 benzyl-analogous cleavage for heteroaromatics γ-cleavage	
Rearrange- ments		for N-heteroaromatics loss of HCN	
		for O-heteroaromatics loss of CO for S-heteroaromatics loss of CS; m/z 45 (CHS$^+$)	
UV		reference data	U130-U155

Characteristic Spectroscopic Data of Halogen Compounds

^{13}C-NMR

Assignment	Range	Comments	Details see page:
alC–F	70–100 ppm	reference data additivity rules (poor for polyhalogenated compounds)	C35,C40 C10
		CF_3: ∿115 ppm	
$\gtrless$C–F	125–175 ppm	additivity rules	C90
C(=C–F)	65–115 ppm		
arC–F	135–165 ppm	additivity rules	C120
arC–(C–F)	105–135 ppm		
		^{19}F–^{13}C coupling constants	C240
alC–Cl	30–60 ppm	reference data additivity rules (poor for polyhalogenated compounds)	C35,C40 C10
$\gtrless$C–Cl	100–150 ppm	additivity rules	C90
arC–Cl	120–150 ppm	additivity rules	C120
alC–Br	10–45 ppm	reference data additivity rules (poor for polyhalogenated compounds)	C35,C40 C10
$\gtrless$C–Br	90–140 ppm	additivity rules	C90
arC–Br	110–140 ppm	additivity rules	C120
alC–I	–20 bis +30 ppm	reference data additivity rules (poor for polyhalogenated compounds)	C35,C40 C10
$\gtrless$C–I	60–110 ppm	additivity rules	C90
arC–I	85–115 ppm	additivity rules	C120

¹H-NMR	$-CH_2-F$	~4.3 ppm	reference data	H5
			^{19}F-^{1}H coupling constants	H355
			additivity rules: olefins	H215
			aromatics	H255
	$-CH_2-Cl$		reference data	H5
	$-CH_2-Br$	~3.5 ppm	aliphatics	H15
			additivity rules: olefins	H215
	$-CH_2-I$	~3.1 ppm	aromatics	H255
IR			reference data	I110
	C–F st	1400–1000 cm^{-1}	strong	
	C–Cl st	<830 cm^{-1}		
	C–Br st	<700 cm^{-1}		
	C–I st	<600 cm^{-1}		
MS	Molecular ion		for saturated aliphatic halogen compounds frequently weak, for polyhalogenated compounds often absent	
			isotope pattern of chlorine and bromine	M100,M105
	Fragments		R–C–hal > R–C–hal	M120
			CF_3: m/z 69; $M^{+\cdot}$-50 or fragment^{+}-50 (CF_2)	
			indications of halogen atoms	M130
	Rearrangements		for fluorine and chlorine compounds –HF or –HCl, respectively	
UV			reference data	U10,U70, U75
	n → σ*	≤280 nm (log ε = ~2.5)	for C–Br and C–Cl generally only endabsorption, for C–F no absorption	

Characteristic Spectroscopic Data of Alcohols and Phenols

	Assignment	Range	Comments	Details see page:
^{13}C-NMR			reference data, additivity rules: aliphatic / alicyclic / aromatic	C10,C30 / C70 / C120,C140
	C-(OH)	50-100 ppm	shift with respect to C-(H) about +40 to +50 ppm	
	C-(C-OH)	10-60 ppm	hardly any shift with respect to C-(C-CH$_3$)	
	C-(C-C-OH)	10-60 ppm	shift with respect to C-(C-C-CH$_3$) about -5 ppm	
	arC-(OH)	135-155 ppm	shift with respect to arC-(H) about +25 ppm, for carbon atoms in ortho- and para-position about -10 ppm	
^{1}H-NMR	H-(O)		reference data	H50
		0.5-5 ppm	aliphatic, alicyclic } position and peak shape strongly dependent on	
		5-8 ppm	aromatic } experimental conditions	
	-CH$_2$-(OH) / >CH-(OH)	3.5-4 ppm	reference data, additivity rules: aliphatic / alicyclic	H15,H55 / H195
	H-(ar-OH)	6.5-7.5 ppm	for C-aromatics shift with respect to H-(ar-H): ortho ∿ -0.6 ppm / meta ∿ -0.1 ppm / para ∿ -0.5 ppm	H255

IR	-OH st	3650–3200 cm^{-1}	reference data band position and shape depends on degree of association	I85
	C–O(H) st	1260–970 cm^{-1}	strong	

MS	Molecular ion		aliphatic: low abundance, often missing for primary or highly branched alcohols; peaks at highest mass are then often due to $M^{+\cdot}-18$ or $M^{+\cdot}-15$ aromatic: abundant	
	Fragments		hints for oxygen	M115,M120 M125
			aliphatic: m/z 31, 45, 59..; $M^{+\cdot}-18$, $M^{+\cdot}-33$, $M^{+\cdot}-46$ primary alcohols: m/z 31 > m/z 45 $\sim$ m/z 59 secondary and tertiary alcohols: local maxima due to α-cleavage: R–CH–R $\xrightarrow{-R\cdot}$ R–CH	
			aromatic: $[ar-O]^{+\cdot}$; $M^{+\cdot}-28$ (CO), $M^{+\cdot}-29$ (CHO), generally accompanied by rearrangement peaks	
	Rearrangements		aliphatic: elimination of H_2O from $M^{+\cdot}$ and products of α-cleavage; elimination of H_2O followed by elimination of olefins	
			olefinic: vinylcarbinols: spectra similar to ketones allyl alcohols: specific, aldehyde elimination:	
			aromatic: if ortho-substitution suitable for:	

$$Y-Z\; : \; -CO-OR, \; C-Hal, \; O-R \; u.\ddot{a}.$$

UV			aliphatic: no absorption above 200 nm	U80,U85
			aromatic: in alkaline solution shift to longer wavelength and more intense	

Characteristic Spectroscopic Data of Ethers

^{13}C-NMR

Assignment	Range	Comments	Details see page:
		reference data, additivity rules: aliphatic / olefinic / aromatic	C10,C40 / C90 / C120
C-(O)	50-100 ppm	for oxiranes outside the normal range	
C-(C-O)		O-C-O: 85-110 ppm	
C-(C-C-O)	10-60 ppm		
(C)=C-(O)	115-165 ppm	shift with respect to (C)=C-(C) about +15 ppm	
C=(C-O)	70-120 ppm	shift with respect to C=(C-C) about -30 ppm	
arC-(O)	135-155 ppm	shift with respect to arC-(H) about +25 ppm, for ortho- or para-carbon atoms about -10 ppm	

^{1}H-NMR

Assignment	Range	Comments	Details see page:
		reference data, additivity rules: aliphatic / cyclic / olefinic / aromatic	H15,H60 / H65 / H215 / H255
CH_3-(O)	3.3-4 ppm	singlet	
$-CH_2$-(O)	3.4-4.2 ppm	$O-CH_2-O$: ∿4.5-6 ppm	
>CH-(O)	3.5-4.3 ppm	$CH-(O)_3$: ∿5-6 ppm	
H-(C<^O_C)	5.7-7.5 ppm	shift with respect to H-(C<^H_C) about +1.2 ppm	
H-(C=C-O)	3.5-5 ppm	shift with respect to H-(C=CH) about -1 ppm	
H-(ar-O)	6.6-7.6 ppm	for C-aromatics shifted with respect to H-(ar-H): ortho ∿ -0.5 ppm / meta ∿ -0.1 ppm / para ∿ -0.4 ppm	

IR				
			reference data	I90
	CH_3-(O) st CH_2-(O) st	2880–2815 cm^{-1}	O-CH_2-O: 2880–2750 cm^{-1}, two bands	
	C-O-C st as	1310–1000 cm^{-1}	strong, sometimes two bands	
MS	Molecular ion		aliphatic: low abundance; tendency to protonate aromatic: abundant	
	Fragments		indications for oxygen	M115,M120
			aliphatic: m/z = 31,45,59,73..; base peak of aliphatic ethers generally due to fragmentation of the bond α to the ether bond: $$[R_1\overset{\alpha}{-}C\text{-}O\text{-}R_2]^{+\cdot} \xrightarrow{-R_1^\cdot} C\overset{+}{=}O\text{-}R_2$$ or due to heterolytic cleavage of the C-O bond, especially for polyethers: $$[R_1\text{-}O\text{-}R_2]^{+\cdot} \xrightarrow{-R_1\text{-}O\cdot} R_2^{+}$$ arylalkyl ethers: preferential loss of the alkyl chain diaryl ethers: preferential loss of CO(28) from $M^{+\cdot}$ and/or $[M\text{-}H]^{+\cdot}$ as well as ar_1-O-ar_2	M125
	Rearrangements		for aliphatic ethers frequently elimination of alcohol for aromatic ethyl and higher ethers:	
UV			aliphatic: no absorption above 200 nm aromatic	
				U80,U105

Characteristic Spectroscopic Data of Amines

^{13}C-NMR

Assignment	Range	Comments	Details see page:
		reference data, additivity rules: aliphatic alicyclic aromatic	C10,C45 C70 C120,C140
C-(N)	25-80 ppm	shift with respect to C-(H) about 20 to 30 ppm	
C-(C-N)	10-60 ppm	shift with respect to C-(C-C) about +2 ppm	
C-(C-C-N)	10-60 ppm	shift with respect to C-(C-C-C) about -2 ppm	
arC-(N)	130-150 ppm	shift with respect to arC-(H) about +20 ppm, for ortho- or para-carbon atoms about -10 to -15 ppm	

^{1}H-NMR

Assignment	Range	Comments	Details see page:
H-(N)		reference data	H75
	0.5-4.0 ppm	aliphatic, alicyclic ⎫ couplings: H80	
	2.5-5.0 ppm	aromatic ⎭	
H-(N$^+$)	6.0-7.0 ppm		
CH$_3$-(N)	2.3-3.1 ppm	reference data, additivity rules: aliphatic	H15,H75
-CH$_2$-(N)	2.5-3.5 ppm	cyclic	H85,H195
		olefinic	H215
>CH-(N)	3.0-3.7 ppm	aromatic	H255
H-(ar-N)	6.0-7.5 ppm	for C-aromatics shifts with respect to H-(ar-H): ortho ∿ -0.8 ppm meta ∿ -0.2 ppm para ∿ -0.7 ppm	H255,H315
H-(C-N$^+$)	3.2-4.0 ppm	couplings: H80	
H-(ar-N$^+$)	7.5-8.0 ppm	for C-aromatics shifts with respect to H-(ar-H): ortho ∿ +0.7 ppm meta ∿ +0.4 ppm para ∿ +0.3 ppm	H255

IR		reference data	I100
	$-NH$ st	$3500-3200\ cm^{-1}$	position of band depends on extent of association
	$-N^+H$ st	$3000-2000\ cm^{-1}$	broad
	$-NH\ \delta$	$1650-1550\ cm^{-1}$	medium
	$-N^+H\ \delta$	$1600-1460\ cm^{-1}$	frequently weak

MS

Molecular ion — odd number if odd number of nitrogens; low abundance; tendency to protonate

Fragments — indications for nitrogen   M115

base peak due to amine cleavage   M125

$$\left[\begin{array}{c} R_1 \\ R_2 \end{array}\!\!\!>\!N\!-\!CH_2\!-\!R_3\right]^{+\cdot} \xrightarrow{-R_3^{\cdot}} \left[\begin{array}{c} R_1 \\ R_2 \end{array}\!\!\!>\!N\!=\!CH_2\right]^{+}$$

fragments of even mass relatively frequent
typical: m/z 30

Rearrangements — elimination of olefins following amine cleavage:

$$\left[\begin{array}{c} R_1 \\ R_2 \end{array}\!\!\!>\!N\!=\!CH_2\right]^{+} \longrightarrow \left[R_1\!-\!NH\!=\!CH_2\right]^{+}$$

elimination of amines from $M^{+\cdot}$:

$$\left[\begin{array}{c} R_1 \\ R_2 \end{array}\!\!\!>\!N\!-\!CH_2\!-\!R_3\right]^{+\cdot} \xrightarrow{-R_1R_2NH} \left[R_3\!-\!CH\right]^{+\cdot}$$

UV

aliphatic: endabsorption   U75

aromatic: in acidic solution shifted to lower wavelength and less intense   U80, U85, U100

<u>Characteristic Spectroscopic Data of Nitro Compounds</u>

^{13}C-NMR

Assignment	Range	Comments	Details see page:
		reference data, additivity rules: aliphatic alicyclic olefinic aromatic	C10,C30 C70 C90 C120,C140
$C-(NO_2)$	55–110 ppm	$C-(C-NO_2)$: shift with respect to $C-(C-C)$ about −6 ppm	
$arC-(NO_2)$	130–150 ppm	frequently broad due to fast spin–spin relaxation; shift with respect to $arC-(H)$ about +20 ppm, for ortho-carbon atoms about −5 ppm, for para-carbon atoms about +6 ppm	

^{1}H-NMR

Assignment	Range	Comments	Details see page:
		reference data, additivity rules: aliphatic alicyclic olefinic aromatic	H15,H90 H195 H215 H255,H315
$H-(C-NO_2)$	4.2–4.6 ppm		
$H-(ar-NO_2)$	7.5–8.5 ppm	for C-aromatics shifted with respect to $H-(ar-H)$: ortho ∿ +1.0 ppm meta ∿ +0.3 ppm para ∿ +0.4 ppm	

IR				
			reference data	I210
	NO_2 st as	$1660-1490$ cm^{-1}	strong to very strong	
	NO_2 st sy	$1390-1260$ cm^{-1}		

MS				
	Molecular ion		aliphatic: low abundance or absent	
			aromatic: abundant; odd numbered if odd number of nitrogens	
	Fragments		indications for nitrogen	M125
			$M^{+\cdot}$ -16, -46	
	Rearrangements		$M^{+\cdot}$ -30, m/z $=$ 30, $M^{+\cdot}$ -17, -47	
			oxygen transfer to substituents in ortho position	

UV				
			for aliphatic nitro compounds weak absorption (log $\varepsilon < 2$) around 275 nm	U15,U60
			aromatic	U50,U75

KOMB

THIOLS, THIOETHERS

Characteristic Spectroscopic Data of Thiols and Thioethers

	Assignment	Range	Comments	Details see page:
^{13}C-NMR			reference data, additivity rules: aliphatic	C10,C30
			alicyclic	C70
			olefinic	C90
			aromatic	C120
	C-(S)	5-60 ppm	no appreciable shift relative to C-(C)	
	arC-(S)	120-140 ppm		
^{1}H-NMR	H-(S-alC)	1.0-2.0 ppm	couplings: H95	
	H-(S-arC)	2.0-4.0 ppm		
	H-(C-S)	2.0-3.2 ppm	reference data, additivity rules: aliphatic	H15,H95
			cyclic	H100,H195
			olefinic	H215
			aromatic	H255
IR			reference data	I215
	-SH st	2600-2540 cm^{-1}	frequently weak	

MS	Molecular ion		stronger than for alcohcls or ethers
			^{34}S-isotope peak; at $M^{+\cdot} + 2 \geqslant 4,5\%$	M60
	Fragments		indications for sulfur	M130
			sulfide cleavage: $[R_1-S-CH_2-R_2]^{+\cdot} \xrightarrow{-R_2^{\cdot}} [R_1-S=CH_2]^+$	
			sulfur peaks: $m/z = 47, 61 \ldots C_nH_{2n+1}S$	M120
	Rearrangements		$M^{+\cdot} -33, -34; m/z = 34, 35, 48$	
			elimination of olefin after sulfide cleavage	
UV	$n \rightarrow \sigma^*$	<225 nm (log ε = 3–4)	in aliphatic compounds	U70,U75
	$n \rightarrow \sigma^*$	220–250 nm (log ε =2–3)		
			aromatic	U95,U105

Characteristic Spectroscopic Data of Aldehydes

	Assignment	Range	Comments	Details see page:
^{13}**C - NMR**	$-C{\lessgtr}^O_H$	190–205 ppm	doublet in partially ("off resonance") decoupled spectra coupling constant $^{13}C-C{\lessgtr}^O_H$: 20–50 Hz, in partially ("off resonance") decoupled spectra usually recognizable	C170
	C-(CHO)	30–70 ppm		C10
	C-(C-CHO)	5–50 ppm	shift with respect to C-(C-CH$_3$) about –10 ppm	
	$\geq$C-(CHO)	110–160 ppm		C90
	C=(C-CHO)			
	ar-(CHO)	120–150 ppm		C120
^{1}**H - NMR**	H-(C=O)	9.0–10.5 ppm	coupling: $H-C-C{\lessgtr}^O_H$ 0–3 Hz $H{\geq}C-C{\lessgtr}^O_H$ $\sim$ 8 Hz	H120
	H-(C-CHO)	2.0–2.5 ppm		H15, H120
	CH=CH-(CHO)	5.5–7.0 ppm		H215
	H-(ar-CHO)	7.2–8.0 ppm	in C-aromatics shifted with respect to H-(ar-H): ortho $\sim$ +0.6 ppm meta $\sim$ +0.2 ppm para $\sim$ +0.3 ppm	

IR	comb C=O st	$2900-2700$ cm^{-1} $1765-1645$ cm^{-1}	reference data two weak bands aliphatic: $\sim$1730 cm^{-1} conjugated: $\sim$1690 cm^{-1}	I120
MS	Molecular ion Fragments Rearrange-ments		aliphatic: moderate aromatic: abundant hints for oxygen $M^{+\cdot} -1$ (for aliphatic aldehydes only to C_7), $M^{+\cdot} -29$ aliphatic: m/z = 44, $M^{+\cdot} -44$ $$\left[R-C\overset{H}{\underset{}{}}O \atop H \right]^{+\cdot} \longrightarrow R-CH=CH_2 + \left[CH_2CHOH \right]^{+\cdot}$$	M125
UV	n → π*	$270-310$ nm (log ε: $\sim$1)	for saturated aldehydes α,β-unsaturated aldehydes benzaldehydes	B65 U20,U65 U40,U95

Characteristic Spectroscopic Data of Ketones

	Assignment	Range	Comments	Details see page:
^{13}C-NMR	$>$C=O	195–220 ppm		C170
	C–(C=O)	25–70 ppm		C10
	C–(C–C=O)	5–50 ppm	shift with respect to C–(C–C) about –6 ppm	
	$>$C–(C=O)	105–160 ppm		C90
	C=(C–C=O)			
	ar–(C=O)	120–150 ppm		C120
^{1}H-NMR	H–(C–C=O)	2.0–3.6 ppm	H–(C–CO–al): 2.0–2.6 ppm H–(C–CO–ar): 2.5–3.6 ppm	H15,H125
	CH=CH–(C=O)	5.5–7.0 ppm		H215
	H–(ar–C=O)	7.2–8.0 ppm	for C-aromatics shifted with respect to H–(ar–H): ortho ⌐ +0.6 ppm meta ⌐ +0.1 ppm para ⌐ +0.2 ppm	H255

IR	C=O st	$1775-1650$ cm^{-1}	reference data
			aliphatic: ~ 1715 cm^{-1}
			cyclic:
			    6-membered and larger rings ~ 1715 cm^{-1}
			    smaller rings > 1750 cm^{-1}
			conjugated: $1690-1665$ cm^{-1}
				I125

MS	Molecular ion		aliphatic: moderate aromatic: abundant	
	Fragments		hints for oxygen ketone cleavages: $$\left[R_1 - \overset{O}{\overset{\|}{C}} - R_2 \right]^{+\cdot} \longrightarrow R_1CO^+,\ R_1^+,\ R_2CO^+,\ R_2^+$$	M115,M120 M125
	Rearrangements		$$\left[\underset{R_2}{R_1 \overset{H}{\underset{O}{}}} \right]^{+\cdot} \xrightarrow{-R_1CH=CH_2} \overset{+}{\underset{}{}}\dot{C}H_2 - \overset{OH}{\overset{\|}{C}} - R_2$$ may occur more than once	
UV	$\pi \to \pi^*$	<200 nm $(\log \varepsilon = 3-4)$	for saturated ketones	U10,U65
	$n \to \pi$	$250-300$ nm $(\log \varepsilon = 1-2)$	α,β-unsaturated ketones phenyl ketones	U20,U60 U40,U90, U105

KOMB

CARBOXYLIC ACIDS

Characteristic Spectroscopic Data of Carboxylic Acids

^{13}C-NMR

Assignment	Range	Comments	Details see page:
$-C\underset{OH}{\overset{O}{\lessgtr}}$	170–185 ppm	in $-COO^-$ shift with respect to $-COOH$ by 0 to +8 ppm	C170
C–(COOH)	20–70 ppm		C10
C–(C–COOH)	5–50 ppm	shift with respect to $C-(C-CH_3)$ about −6 ppm	
$\gtrless$C–(COOH)	105–160 ppm		C90
C=(C–COOH)			
ar–(COOH)	120–150 ppm		C120

^{1}H-NMR

Assignment	Range	Comments	Details see page:
H–(O–CO)	10.0–13.0 ppm	position and peak shape strongly dependent on experimental conditions	H135
H–(C–COOH)	2.0–2.6 ppm		H15,H135
CH=CH–(COOH)	5.2–7.5 ppm	shift with respect to CH=CH–(H) : geminal ∿ +0.9 ppm cis ∿ +1.0 bis +1.4 ppm trans ∿ +0.3 bis +0.7 ppm	H215
H–(ar–COOH)	7.5–8.5 ppm	for C-aromatics shifted with respect to H–(ar–H) : ortho ∿ +0.8 ppm meta ∿ +0.2 ppm para ∿ +0.3 ppm	H255

			reference data	I165
IR	COO–H st	3550–2500 cm^{-1}	broad	
	C=O st	1800–1650 cm^{-1}	aliphatic: $\sim$1715 cm^{-1} conjugated: $\sim$1695 cm^{-1} in COO$^-$ two bands at 1580 cm^{-1} and 1420 cm^{-1}	
	CO–OH δ oop	$\sim$920 cm^{-1}		
MS	Molecular ion		aliphatic: moderate, for long chains more abundant; tendency to protonate aromatic: abundant	
	Fragments		hints for oxygen $M^{+\cdot}$ –17 (for aromatic carboxylic acids abundant), $M^{+\cdot}$ –45	M115 M125
	Rearrangements		aliphatic: $M^{+\cdot}$ –18, m/z = 60,61 aromatic: ortho-effect:	

UV	n → π*	<220 nm (log ε = 1–2)	for saturated acids	U10
			α,β-unsaturated acids	U20,U60, U65
			benzoic acids	U40,U90, U95

KOMB

Characteristic Spectroscopic Data of Carboxylic Esters and Lactones

^{13}**C-NMR**

Assignment	Range	Comments	Details see page:
$-C\lessgtr^{O}_{O-}$	165–180 ppm	shift with respect to -COOH by -5 to -10 ppm	C170
C-(COO-)	20–70 ppm		C10
C-(C-COO-)	5–50 ppm	shift with respect to C-(C-C) about -6 ppm	
C-(OOC-)	50–100 ppm	shift with respect to C-(OH) about +2 to +10 ppm	
C-(C-OOC-)	10–60 ppm	shift with respect to C-(C-OH) about -4 ppm	
$\geq$C-(COO-) C=(C-COO-)	105–160 ppm		C90
$\geq$C-(OOC-)	100–150 ppm		
C=(C-OOC-)	80–130 ppm		
ar-(COO-)	120–150 ppm		C120
ar-(OOC-)	100–160 ppm	shift with respect to ar-(H): α ~ +20 ppm ortho ~ - 6 ppm meta ~ + 1 ppm para ~ - 2 ppm	

^{1}H-NMR

		reference data, additivity rules: aliphatic olefinic aromatic	H15,H140 H215 H255
H-(C-COO-)	2.0-2.5 ppm	CH_3COO-: ∿2.0 ppm; $-CH_2COO-$: ∿2.3 ppm; $>$CHCOO-: ∿2.5 ppm	H140
H-(C-OOC-)	3.5-5.3 ppm	CH_3OOC-: 3.5-3.9 ppm; $-CH_2OOC-$: 4.0-4.5 ppm; $>$CHOOC-: 4.8-5.3 ppm	
CH=CH-(COO-)	5.2-7.5 ppm	shift with respect to CH=CH-(H): geminal ∿ +0.8 ppm cis ∿ +1.1 ppm trans ∿ +0.5 ppm	H215
$=C<^H_{(OOC-)}$	6.0-8.0 ppm	shift with respect to CH=CH-(H): geminal ∿ +2.1 ppm cis ∿ -0.4 ppm trans ∿ -0.6 ppm	
HC=(C-OOC-)	4.5-6.0 ppm		
H-(ar-COO-)	7.5-8.5 ppm	in C-aromatics shift with respect to H-(ar-H): ortho ∿ +0.7 ppm meta ∿ +0.1 ppm para ∿ +0.2 ppm	H255
H-(ar-OOC-)	6.8-7.5 ppm	in C-aromatics shift with respect to H-(ar-H): ortho ∿ -0.2 ppm meta ∿ 0 ppm para ∿ -0.1 ppm	

Characteristic Spectroscopic Data of Carboxylic Esters and Lactones
(continued)

	Assignment	Range	Comments	Details see page:
IR			reference data	I135
	C=O st	$1745-1730\ cm^{-1}$	strong; quoted range holds for aliphatic esters	
			shifted to higher wavenumbers for hal-C-COO-, -COO-C=C, -COO-ar and for small-ring lactones	
			shifted to lower wavenumbers for C=C-COO- and ar-COO-	
	C-O st	$1330\dot{-}1050\ cm^{-1}$	two bands, at least one of them strong	
MS	Molecular ion		esters: aliphatic: low abundance, tendency to protonate	
			aromatic: abundant	
			lactones: aliphatic: medium to low abundance, tendency to protonate	
			aromatic: strong	
	Fragments		indications for oxygen	M115
			esters: $M^{+\cdot}$ - RO, $M^{+\cdot}$ - ROCO	M125
			lactones: loss of α-substituents (attached to ether-carbon), decarbonylation, for aromatic lactones double decarbonylation	

Rearrangements				U10 U20 U30 U40
			esters: elimination of olefin from the alcohol moiety: $$\left[R_1-\overset{\overset{O}{\|}}{C}\overset{H}{\underset{O}{\diagdown}}R_2 \right]^{+\cdot} \longrightarrow \left[R_1-COOH + R_2-CH=CH_2 \right]^{+\cdot}$$ elimination of the alcohol side chain with transfer of two hydrogens: $$\left[R_1COOR_2 \right]^{+\cdot} \longrightarrow R_1C\overset{+}{\underset{OH}{\diagup}}{}^{OH}$$ aliphatic esters: elimination of the alkyl chain of the acid moiety as an olefin: $$\left[R_1\cdots O \cdots OR_2 \right]^{+\cdot} \xrightarrow{-\ R_1-CH=CH_2} \cdot CH_2-\overset{\overset{+}{OH}}{\underset{\|}{C}}-OR_2$$ aromatic esters: ortho elimination: $$\left[\text{(aryl)}\,\overset{O}{\underset{OR}{C}}\cdots X-H \right]^{+\cdot} \xrightarrow{-\ ROH} \left[\text{(aryl)}\,C=O \cdots X \right]^{+\cdot}$$ **lactones:** $M^{+\cdot} - 18$
UV	$n \to \pi^*$	<220 nm ($\log \varepsilon = 1\text{--}2$)	for aliphatic esters $C=C-COO-$ $COO-C=C-C=C$ (phenyl)$-COO-$	U10 U20 U30 U40

KOMB

AMIDES, LACTAMS

<u>Characteristic Spectroscopic Data of Amides and Lactams</u>

	Assignment	Range	Comments	Details see page:
^{13}**C-NMR**	$C{\overset{\nearrow O}{\underset{\searrow N}{}}}$	165-180 ppm		C170
	C-(CON)	20-70 ppm		C10
	C-(C-CON)	5-50 ppm	shift with respect to C-(C-C) about -6 ppm	
	C-(NCO)	25-80 ppm	shift with respect to C-(NH) about -1 to -2 ppm	C10
	C-(C-NCO)	10-60 ppm		
	ar-(CON)	120-150 ppm		C120
	ar-(NCO)	110-150 ppm	shift with respect to ar-(H): α ～ +11 ppm ortho ～ -10 ppm meta ～ 0 ppm para ～ - 6 ppm	
^{1}**H-NMR**	H-(NCO)	5-10 ppm	frequently broad to very broad; splitting due to H-N-C-H coupling often only recognizable in the CH-signal, see H150	H150
	H-(C-CON)	2.0-2.5 ppm		H5
	H-(C-NCO)	2.7-4.8 ppm	CH_3-NCO 2.7-3.0 ppm -CH_2-NCO 3.1-3.5 ppm CH-NCO 3.8-4.8 ppm	H155

CH=CH-(CON)	5.2-7.5 ppm	shift with respect to CH=CH-(H): geminal $\sim$ +1.4 ppm cis $\sim$ +1.0 ppm trans $\sim$ +0.5 ppm	H215
=C$<^H_{NCO}$ HC=(C-NCO)	6.0-8.0 ppm 4.5-6.0 ppm	shift with respect to CH=CH-(H): geminal $\sim$ +2.1 ppm cis $\sim$ -0.6 ppm trans $\sim$ -0.7 ppm	
H-(ar-CON)	7.5-8.5 ppm	for C-aromatics shift with respect to H-(ar-H): ortho $\sim$ +0.6 ppm meta $\sim$ +0.1 ppm para $\sim$ +0.2 ppm	H255
H-(ar-NCO)	6.8-7.5 ppm	for C-aromatics shift with respect to H-(ar-H): ortho $\sim$ 0 ppm meta $\sim$ 0 ppm para 0 to -0.3 ppm	
		reference data	I145
N-H st	3500-3100 cm^{-1}	position and shape of band depends on degree of association; frequently multiplicity of bands	
C=O st (amide I)	1700-1650 cm^{-1}	strong; range listed holds for amides as well as δ- and larger lactams for β- and γ-lactams higher	
N-H δ N-C=O st sy (amide II)	1630-1510 cm^{-1}	frequently strong; missing in tertiary amides and in lactams	

IR

Characteristic Spectroscopic Data of Amides and Lactams

(continued)

	Assignment	Range	Comments	Details see page:
MS	Molecular ion		aliphatic: moderate abundance, tendency to protonate aromatic: abundant	
	Fragments		aliphatic amides indications for nitrogen <u>amides:</u> amide cleavage resulting in acyl ion, followed by loss of CO; unusually large number of fragments of even mass <u>lactams:</u> loss of α-substituents, loss of CO	M120 M125
	Rearrangements		<u>amides:</u> elimination of the amine moiety elimination of an olefin from the amine or acid moiety in analogy to esters (see B230) <u>lactams:</u> $M^{+\cdot} - 18$	
UV	n → π*	<220 nm (log ε = 1-2)	for aliphatic amides and lactams aromatic	U10 U85

^{13}C-Chemical Shifts in Alkanes

(δ in ppm relative to TMS)

CH_4 $-$ 2.3

$$\begin{array}{l} CH_3 \\ | \\ CH_3 \end{array} \quad 8.4$$

$$\begin{array}{l} CH_3 \quad 15.4 \\ | \\ CH_2 \quad 15.9 \\ | \\ CH_3 \end{array}$$

$$\begin{array}{l} CH_3 \quad 13.0 \\ | \\ CH_2 \quad 24.8 \\ | \\ CH_2 \\ | \\ CH_3 \end{array}$$

$$\begin{array}{l} CH_3 \quad 24.1 \\ | \\ CH \quad 25.0 \\ | \\ (CH_3)_2 \end{array}$$

$$\begin{array}{l} CH_3 \quad 13.5 \\ | \\ CH_2 \quad 22.4 \\ | \\ CH_2 \quad 34.3 \\ | \\ CH_2 \\ | \\ CH_3 \end{array}$$

$$\begin{array}{l} CH_3 \quad 11.3 \\ | \\ CH_2 \quad 31.6 \\ | \\ CH \quad 29.7 \\ | \\ (CH_3)_2 \quad 21.8 \end{array}$$

$$\begin{array}{l} CH_3 \quad 31.3 \\ | \\ C \quad 27.7 \\ | \\ (CH_3)_3 \end{array}$$

The ^{13}C-chemical shifts in alkanes can be estimated using the additivity rule stated on p. C10.

For cycloalkanes see p. C50.

^{13}C-NMR

ALIPHATICS, ADDITIVITY RULE

Estimation of the ^{13}C-Chemical Shifts in Aliphatic Compounds

(δ in ppm relative to TMS)

$$\delta = -2.3 + \sum_i Z_i + S + \sum_j K_j$$

Substituent		Increment Z_i for substituents in position				Additional data see page:
		α	β	γ	δ	
C	-H	0.0	0.0	0.0	0.0	
	-C≡ (*)	9.1	9.4	-2.5	0.3	C5
	-C△O (*)	21.4	2.8	-2.5	0.3	C40
	-C=C- (*)	19.5	6.9	-2.1	0.4	C85–C95
	-C≡C-	4.4	5.6	-3.4	-0.6	C110
	-phenyl	22.1	9.3	-2.6	0.3	C115
HAL	-F	70.1	7.8	-6.8	0.0	C40,C240
	-Cl	31.0	10.0	-5.1	-0.5	C40
	-Br	18.9	11.0	-3.8	-0.7	C40
	-I	-7.2	10.9	-1.5	-0.9	C40
O	-O- (*)	49.0	10.1	-6.2	0.0	C40
	-O-CO-	56.5	6.5	-6.0	0.0	C175
	-O-NO	54.3	6.1	-6.5	-0.5	
N	-N< (*)	28.3	11.3	-5.1	0.0	C45
	-N$^+$≡ (*)	30.7	5.4	-7.2	-1.4	
	-NH$_3^+$	26.0	7.5	-4.6	0.0	
	-NO$_2$	61.6	3.1	-4.6	-1.0	
	-NC	31.5	7.6	-3.0	0.0	C190
S	-S- (*)	10.6	11.4	-3.6	-0.4	
	-S-CO-	17.0	6.5	-3.1	0.0	C200
	-SO- (*)	31.1	9.0	-3.5	0.0	
	-SO$_2$Cl	54.5	3.4	-3.0	0.0	
	-SCN	23.0	9.7	-3.0	0.0	
O=C<	-CHO	29.9	-0.6	-2.7	0.0	C170
	-CO-	22.5	3.0	-3.0	0.0	C170
	-COOH	20.1	2.0	-2.8	0.0	C170
	-COO$^-$	24.5	3.5	-2.5	0.0	C170
	-COO-	22.6	2.0	-2.8	0.0	C170
	-CON<	22.0	2.6	-3.2	-0.4	C170
	-COCl	33.1	2.3	-3.6	0.0	C170
	-CS-N<	33.1	7.7	-2.5	0.6	C200
	-C=NOH syn	11.7	0.6	-1.8	0.0	C195
	-C=NOH anti	16.1	4.3	-1.5	0.0	C195
	-CN	3.1	2.4	-3.3	-0.5	C190
	-Sn<	-5.2	4.0	-0.3	0.0	

Steric Corrections S

Observed ^{13}C-center	Number of substituents other than H at the most branched α-substituent (not for those α-substituents marked by (*) on p. C10)			
	1	2	3	4
primary	0.0	0.0	-1.1	-3.4
secondary	0.0	0.0	-2.5	-7.5
tertiary	0.0	-3.7	-9.5	-15.0
quaternary	-1.5	-8.4	-15.0	-25.0

Conformation Corrections K for γ-Substituents

conformation		K
synperiplanar		-4.0
synclinal		-1.0
anticlinal		0.0
antiperiplanar		2.0
not fixed	—	0.0

^{13}C-NMR

Example: Estimation of the ^{13}C-chemical shifts for
N-t-butoxycarbonylalanine

$$
\begin{array}{c}
\overset{\textbf{b}}{CH_3} \\
| \\
\underset{\textbf{d}}{(CH_3)_3}\text{-}\underset{\textbf{c}}{C\text{-}O\text{-}CO}\text{-}NH\text{-}\underset{\textbf{a}}{CH}\text{-}COOH
\end{array}
$$

(a) base value:	-2.3		(b) base value:	-2.3
1αC	9.1		1αC	9.1
1αCOOH	20.1		1βCOOH	2.0
1αNH	28.3		1βNH	11.3
1βCOO	2.0		1γCOO	-2.8
1δC	0.3		S(p,3)	-1.1
S(t,2)	-3.7			
			estimated:	16.2
estimated:	53.8		determined:	17.3
determined:	49.0			

(c) base value:	-2.3		(d) base value:	-2.3
3αC	27.3		1αC	9.1
1αOCO	56.5		2βC	18.8
1γNH	-5.1		1βOCO	6.5
1δC	0.3		1δNH	0.0
S(q,1)	-1.5		S(p,4)	-3.4
estimated:	75.2		estimated:	28.7
determined:	78.1		determined:	28.1

The ^{13}C-chemical shifts estimated for aliphatic hydrocarbons on the
basis of the additivity rule differ in general by less than about
5 ppm from the experimental values. Larger discrepancies may be
expected for highly branched systems (particularly for quaternary
carbon atoms). For chemical shifts exceeding about 90 ppm the de-
viations can be so large as to render the rule useless.

One can also use the chemical shift of a reference compound as the base value if its structure is closely related to that assumed for the unknown. The increments corresponding to the structural elements missing in the reference compound are then added to the base value, while those of structural elements present in the reference but absent in the unknown are subtracted.

Example: Estimation of the ^{13}C-chemical shifts for (a) and (b) in N-t-butoxycarbonylalanine (see p. C20) using the chemical shifts of valine as base values (a', b').

$$(CH_3)_2\text{-CH-CH-COOH}$$

with NH$_2$ on the second carbon (a'), labelled b' (CH) and a' (CH-COOH).

(a')   61.6

(b')   30.2

(a) base value:	61.6	(b) base value:	30.2
+1βCOO	2.0	+1γCOO	-2.8
+1δC	0.3	+S(p,3)	-1.1
+S(t,2)	-3.7		
-2βC	-18.8	-2αC	-18.2
-S(t,3)	9.5	-S(t,3)	9.5
estimated:	50.9	estimated:	17.6
determined:	49.0	determined:	17.3

^{13}C-NMR

^{13}C-Chemical Shifts for Methyl Groups

(δ in ppm relative to TMS)

	Substituent X	δCH$_3$-X
C	-H	-2.3
	-CH$_3$	8.4
	-CH$_2$CH$_3$	15.4
	-CH(CH$_3$)$_2$	24.1
	-C(CH$_3$)$_3$	31.3
	-(CH$_2$)$_6$CH$_3$	14.1
	-CH$_2$-phenyl	15.7
	-CH$_2$F	14.4
	-CH$_2$Cl	17.7
	-CH$_2$Br	20.2
	-CH$_2$I	23.0
	-CH$_2$OH	18.8
	-CH$_2$OCOC$_8$H$_{17}$	14.3
	-CH$_2$OCH$_3$	15.9
	-CH$_2$CHO	5.2
	-CH$_2$COCH$_3$	7.3
	-CH$_2$COOH	9.0
	-cyclopentyl	20.5
	-cyclohexyl	23.1
	-phenyl	21.4
	-α-naphthyl	19.1
	-β-naphthyl	21.5
	-2-pyridyl	24.2
	-3-pyridyl	18.0
	-4-pyridyl	20.6
	-2-furyl	13.7
	-2-thienyl	14.7
	-1-pyrrolyl	35.9
	-2-pyrrolyl	11.8
	-1-imidazolyl	32.2
	-1-pyrazolyl	38.4
	-1-indolyl	32.1
	-2-indolyl	13.4
	-3-indolyl	9.8
	-4-indolyl	21.6
	-5-indolyl	21.5
	-6-indolyl	21.7
	-7-indolyl	16.6

	Substituent X	δCH$_3$-X
HAL	-F	75.2
	-Cl	24.9
	-Br	10.0
	-I	-20.7
O	-OH	50.2
	-OCH$_3$	60.9
	-OCH$_2$CH$_3$	58.8
	-OCH(CH$_3$)$_2$	56.1
	-O-cyclohexyl	55.1
	-O-phenyl	54.0
	-OCOC$_8$H$_{17}$	51.4
	-OCO-cyclohexyl	51.0
	-OCOCH=CH$_2$	50.9
N	-NH$_2$	26.9
	-NHCH$_3$	
	-N(CH$_3$)$_2$	47.5
	-NH-cyclohexyl	33.5
	-NH-phenyl	30.2
	-N(CH$_3$)-phenyl	39.9
	-N(CH$_3$)-CHO	36.2;31.1
	-NO$_2$	57.1
	-NC	26.8
S	-SCH$_3$	19.3
	-SC$_8$H$_{17}$	15.5
	-S-phenyl	15.6
	-SOCH$_3$	43.3
O=C	-CHO	31.2
	-COCH$_3$	28.1
	-CO-cyclohexyl	27.6
	-COCH=CH$_2$	
	-CO-phenyl	24.9
	-COOH	21.1
	-COOCH$_3$	20.0
	-COSC$_4$H$_9$	30.1
	-CON(C$_4$H$_9$)$_2$	21.4
	-CN	1.3

^{13}C-Chemical Shifts in Substituted n-Octanes

(δ in ppm relative to TMS)

Substituent X	X-CH$_2$	-CH$_2$	-CH$_2$	-CH$_2$	-CH$_2$	-CH$_2$	-CH$_2$	-CH$_3$
-H	14.1	22.8	32.1	29.5	29.5	32.1	22.8	14.1
-CH=CH$_2$	34.5	~29.6	~29.6	~29.6	~29.6	32.2	23.0	13.9
-phenyl	36.2	31.7	~29.6	~29.6	~29.6	32.1	22.8	14.1
-F[1)	84.2	30.6	25.3	29.3	29.3	31.9	22.7	14.1
-Cl	45.1	32.8	27.0	29.0	29.2	31.9	22.8	14.1
-Br	33.8	33.0	28.3	28.8	29.2	31.8	22.7	14.1
-I	6.9	33.7	30.6	28.6	29.1	31.8	22.6	14.1
-OH	63.1	32.9	25.9	29.5	29.4	31.9	22.8	14.1
-OC$_8$H$_{17}$	71.1	30.0	26.3	29.6	29.4	32.0	22.8	14.1
-ONO	68.3	29.2	26.0	29.3	29.3	31.9	22.7	14.0
-NH$_2$	42.4	34.1	27.0	29.6	29.4	31.9	22.7	14.1
-NO$_2$	75.8	26.2	27.9	~29.6	~29.6	31.4	22.6	14.0
-SH	24.7	34.2	28.5	29.2	29.1	31.9	22.7	14.1
-SCH$_3$	34.5	29.0	29.4	29.4	29.4	31.9	22.8	14.1
-SOC$_8$H$_{17}$	52.6	~29.1	~29.1	~29.1	~29.1	31.8	22.7	14.1
-CHO	44.0	22.2	~29.3	~29.3	~29.3	31.9	22.7	14.1
-COCH$_3$	43.7	24.1	~29.5	~29.5	~29.5	32.0	22.8	14.1
-COOH	34.2	24.8	~29.3	~29.3	~29.3	31.9	22.7	14.1
-COOCH$_3$	34.2	25.1	29.3	29.3	29.3	31.9	22.8	14.1
-COCl	47.2	25.1	28.5	29.1	29.1	31.8	22.7	14.1
-CN	17.2	25.5	~29.9	~29.9	~29.9	31.8	22.7	14.0

The left margin groups the substituents: **C**, **H**, **A**, **L**, **O**, **N**, **S** (spelling CHALONS) and an $\overset{O}{\underset{\diagdown}{\overset{\parallel}{C}}}$ group.

1) $|J|_{CF} = 164.8$ Hz, $\quad |J|_{C-CF} = 18.3$ Hz, $\quad |J|_{C-C-CF} = 6.2$ Hz,

$|J|_{C-C-C-C-F} = 0$ Hz

HALOMETHANES, CYCLIC ETHERS

^{13}C-Chemical Shifts in Halomethanes

(δ in ppm relative to TMS)

X	CH_3X	CH_2X_2	CHX_3	CX_4
F	75.2	109.0	116.4	118.5
Cl	24.9	54.0	77.2	96.1
Br	10.2	21.4	12.1	-28.7
I	-20.7	-54.0	-139.9	-292.5

^{13}C-Chemical Shifts in Cyclic Ethers

(δ in ppm relative to TMS)

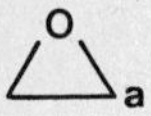

(a) 39.5

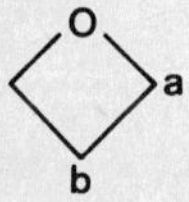

(a) 72.6
(b) 22.9

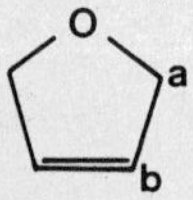

(a) 75.3
(b) 126.3

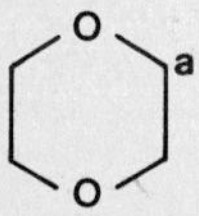

(a) 67.6

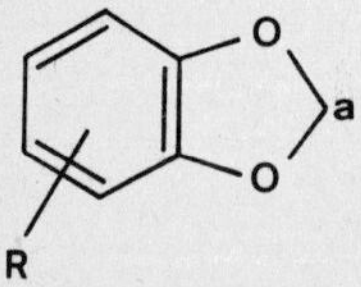

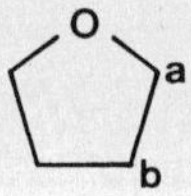

(a) 68.4
(b) 26.5

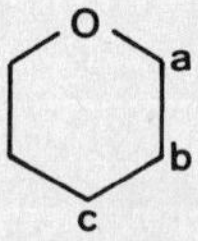

(a) 69.5
(b) 27.7
(c) 24.9

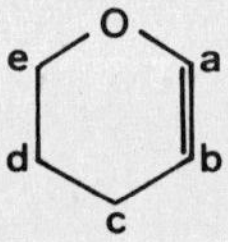

(a) 144.1
(b) 99.4
(c) 19.4
(d) 22.6
(e) 64.8

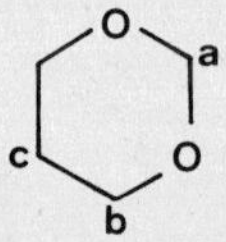

(a) 94.8
(b) 67.5
(c) 27.5

(a) ~101

¹³C-Chemical Shifts in Cyclic Amines

(δ in ppm relative to TMS)

(a) 47.1
(b) 25.7

(a) 47.9

(a) 47.9
(b) 27.8
(c) 25.9

(a) 48.6
(b) 28.5

(a) 47.7
(b) 57.2
(c) 26.4
(d) 26.4

(a) 46.4
(b) 57.7
(c) 17.5

(a) 45.9
(b) 54.2
(c) 125.0
(d) 124.3
(e) 26.2
(f) 51.7

(a) 42.7
(b) 56.7
(c) 24.4

(a) 48.0
(b) 58.9
(c) 28.5
(d) 28.5

^{13}C-NMR

<u>^{13}C-Chemical Shifts in Saturated Alicycles</u>

(δ in ppm relative to TMS)

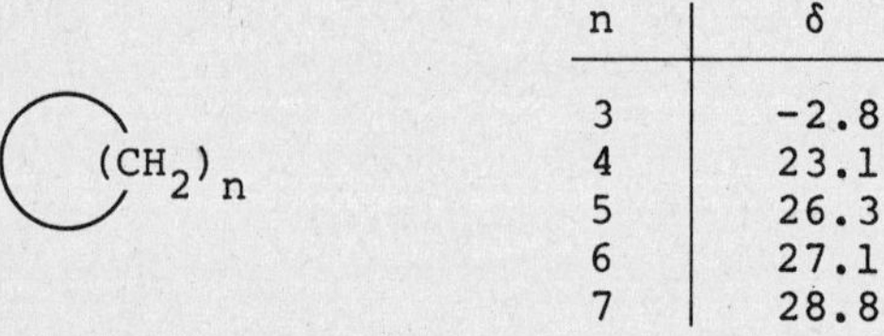

n	δ
3	-2.8
4	23.1
5	26.3
6	27.1
7	28.8

Unsaturated alicycles: see p. C100.

Condensed alicycles: see p. C75.

<u>Additivity Rule for Estimating ^{13}C-Chemical Shifts for Methyl Groups</u>
<u>in Methylcyclohexanes</u>

(δ in ppm relative to TMS)

base value: 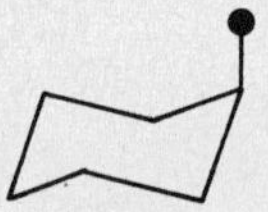 18.8    23.1

increments for
methyl substituents:
(example: see p. C60)

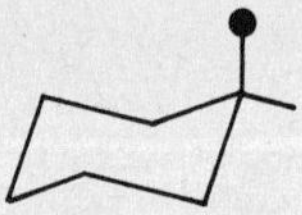

α	6.4	10.4
βeq	-6.8	-2.8
βax	-2.8	-2.8
γax	2.0	0.0
γeq	0.0	0.0
δax, δeq	0.0	0.0

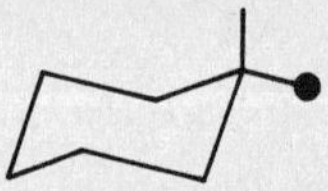

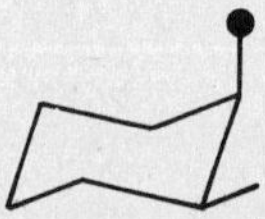

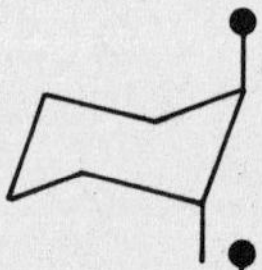

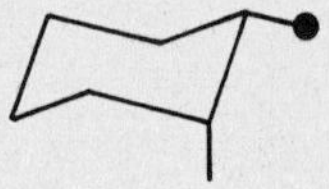

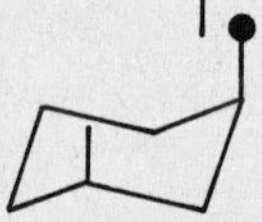

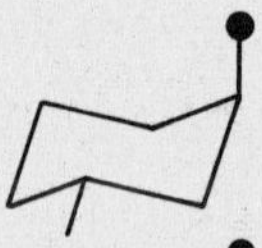

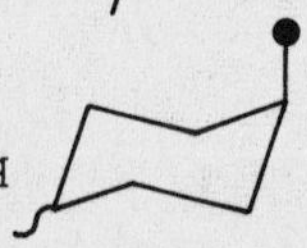

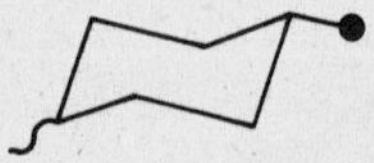

<u>Additivity Rule for Estimating ^{13}C-Chemical Shifts for Ring Carbons in Methyl-Substituted Cyclohexanes</u>

(δ in ppm relative to TMS)

base value: 27.1

increments:

- for equatorial C-substituents			- for axial C-substituents		
αeq		6.0	αax		1.4
βeq		9.0	βax		5.4
γeq		0.0	γax		-6.4
δeq		-0.2	δax		-0.1

correction terms:

- for geminal disubstitution			- for vicinal disubstitution		
αax,αeq		-3.8	αeq,βeq		-2.5
βax,βeq		-1.3	αeq,βax		-2.9
γax,γeq		2.0	αax,βeq		-3.4
			βeq,γax		-0.8
			βax,γeq		1.6

^{13}C-NMR

ALICYCLICS, ADDITIVITY RULE

<u>Example</u>: Estimation of the ^{13}C-chemical shifts in
1-cis-2-cis-4-trimethylcyclohexane:

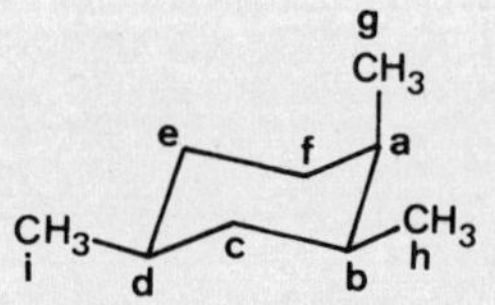

(a)				(b)		
	base value:	27.1			base value:	27.1
	1αax	1.4			1αeq	6.0
	1βeq	9.0			1βax	5.4
	1δeq	-0.2			1γeq	0.0
	1αax,βeq	-3.4			1αeq,βax	-2.9
	estimated:	33.9			estimated:	35.6
	determined:	33.7 (34.1)			determined:	35.5

(c)				(d)		
	base value:	27.1			base value:	27.1
	2βeq	18.0			1αeq	6.0
	1γax	-6.4			1γeq	0.0
	1βeq,γax	-0.8			1δax	-0.1
	estimated:	37.9			estimated:	33.0
	determined:	38.0			determined:	32.9

(e)				(f)		
	base value:	27.1			base value:	27.1
	1βeq	9.0			1βax	5.4
	1γax	-6.4			2γeq	0.0
	1δeq	-0.2			1βax,γeq	1.6
	estimated:	29.5			estimated:	34.1
	determined:	29.3			determined:	33.7 (34.1)

<table>
<tr><td>(g)</td><td>base value:</td><td>18.8</td><td></td><td>(h)</td><td>base value:</td><td>23.1</td></tr>
<tr><td></td><td>1CH$_3$βeq</td><td>-6.8</td><td></td><td></td><td>1CH$_3$βax</td><td>-2.8</td></tr>
<tr><td></td><td>1CH$_3$δeq</td><td>0.0</td><td></td><td></td><td>1CH$_3$γeq</td><td>0.0</td></tr>
<tr><td></td><td>estimated:</td><td>12.0</td><td></td><td></td><td>estimated:</td><td>20.3</td></tr>
<tr><td></td><td>determined:</td><td>11.7</td><td></td><td></td><td>determined:</td><td>20.3</td></tr>
</table>

<table>
<tr><td>(i)</td><td>base value:</td><td>23.1</td></tr>
<tr><td></td><td>1CH$_3$γeq</td><td>0.0</td></tr>
<tr><td></td><td>1CH$_3$δax</td><td>0.0</td></tr>
<tr><td></td><td>estimated:</td><td>23.1</td></tr>
<tr><td></td><td>determined:</td><td>23.0</td></tr>
</table>

The ^{13}C-chemical shifts in substituted alicyclics can be estimated
using the additivity rule for aliphatic hydrocarbons (see p. C10).
For this it is advisable to take as base value the chemical shifts
for the unsubstituted (or similarly substituted) alicyclic compound.
The discrepancies between experimental and calculated values are
generally somewhat larger than those found for aliphatic compounds
(see p. C20).

^{13}C-NMR

SUBSTITUTED CYCLOHEXANES

<u>^{13}C-Chemical Shifts in Monosubstituted Cyclohexanes</u>

(δ in ppm relative to TMS)

Substituent X	^{13}C-chemical shift for			
	a	b	c	d
-H	27.6	27.6	27.6	27.6
$-CH_3$	33.4	36.0	27.1	27.0
$-CH_2CH_3$	40.2	33.7	27.1	27.4
$-CH_2CH_2CH_2CH_3$	38.4	34.1	27.1	27.3
$-C(CH_3)_3$	48.8	28.1	27.7	27.1
-cyclohexyl	44.3	30.8	27.4	27.4
-phenyl	45.1	34.9	27.4	26.7
$-F^{1)}$	90.5	33.1	23.5	26.0
-Cl	59.8	37.2	25.2	25.6
-Br	52.6	37.9	26.1	25.6
-I	31.8	39.8	27.4	25.5
-OH	70.0	36.0	25.0	26.4
$-OCH_3$	78.6	32.3	24.3	26.7
$-OCOCH_3$	72.3	32.2	24.4	26.1
$-NH_2$	51.1	37.7	25.8	26.5
$-NH_3^+Cl^-$	51.5	33.4	25.6	26.0
-N=C=N-cyclohexyl	55.7	35.0	24.7	25.5
$-NO_2$	84.6	31.4	24.7	25.5
-SH	38.5	38.5	26.8	25.9
$-COCH_3$	51.5	29.0	26.6	26.3
-COOH	43.7	29.6	26.2	26.6
$-COO^-$	47.2	30.9	26.9	26.9
$-COOCH_3$	43.4	29.6	26.0	26.4
-COCl	55.4	29.7	25.5	25.9
-CN	28.3	30.1	24.6	25.8

The row labels in the left margin group the substituents as: **C** (carbon), **HAL** (halogens), **O**, **N**, **S**, **O=C** (carbonyl).

$^{1)}$ $|J|_{CF} = 171$ Hz, $|J|_{C-CF} = 19$ Hz, $|J|_{C-C-CF} = 5$ Hz

C70

^{13}C-Chemical Shifts in Some Saturated Condensed Alicyclics

(δ in ppm relative to TMS)

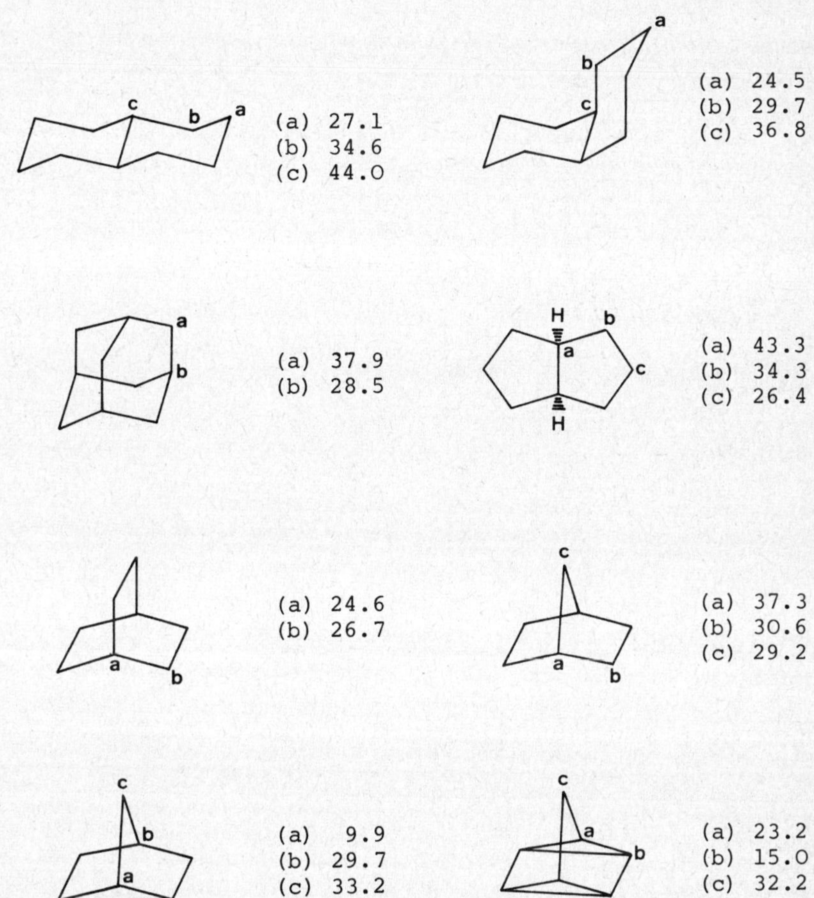

Unsaturated condensed alicyclics: see p. C100

Alkenes

The ^{13}C-chemical shifts of the carbons of C=C double bonds range from ca. 80 - 160 ppm. In <u>unsubstituted alkenes</u> they can be estimated quite accurately (see p. C85).

To estimate the ^{13}C-chemical shifts in <u>substituted alkenes</u> one can use the substituent effects listed for the ^{13}C-chemical shifts in vinyl groups (see the example on p. C95).

The ^{13}C-chemical shifts of sp^3-hybridized carbons in the vicinity of double bonds can be estimated using the additivity rule outlined on p. C10. The conformational correction factor K differs widely for γ-substituents of cis *vs.* trans-disubstituted olefins because the conformation is fixed by the double bond. It is thus quite easy to assign the correct isomeric structure.

Estimation of the ^{13}C-Chemical Shifts of sp^2-Hybridized Carbons in <u>Unsubstituted Alkenes</u> (δ in ppm relative to TMS)

$$C-C-C-C'=C-C-C-C$$
$$\gamma'\ \beta'\ \alpha' \qquad \uparrow\ \alpha\ \beta\ \gamma$$

base value: 123.3

increments for C-substituents:

- of the C-atom under consideration (C)		- at the neighboring C-Atom (C')	
α	10.6	α'	-7.9
β	4.9	β'	-1.8
γ	-1.5	γ'	1.5

steric corrections:

- for each pair of cis α,α'-substituents: -1.1
- for a pair of geminal α,α-substituents: -4.8
- for a pair of geminal α',α'-substituents: 2.5
- if one or more β-substituents are present: 2.3

<u>Example:</u> Estimation of the chemical shifts in cis-4-methyl-2-pentene:

$$CH_3-CH{=}CH-CH{<}^{CH_3}_{CH_3}$$
a   b

(a)			(b)		
base value:	123.3		base value:	123.3	
1αC	10.6		1αC	10.6	
1α'C	-7.9		2βC	9.8	
2β'C	-3.6		1α'C	-7.9	
correction:			correction:		
cis α,α'	-1.1		cis α,α'	-1.1	
			1β-substituent	2.3	
estimated:	121.3		estimated:	137.0	
determined:	121.8		determined:	138.8	

SUBSTITUTED ALKENES

Effect of a Substituent on the ^{13}C-Chemical Shifts in Vinyl Compounds (δ in ppm relative to TMS)

$$\underset{1 \quad 2}{X{-}CH{=}CH_2} \qquad \delta_{C_i} = 123.3 + z_i$$

	Substituent X	z_1	z_2
C	$-H$	0.0	0.0
	$-CH_3$	10.6	-7.9
	$-CH_2CH_3$	15.5	-9.7
	$-CH_2CH_2CH_3$	14.0	-8.2
	$-CH(CH_3)_2$	20.4	-11.5
	$-CH_2CH_2CH_2CH_3$	14.6	-8.9
	$-C(CH_3)_3$	25.3	-13.3
	$-CH_2Cl$	10.2	-6.0
	$-CH_2Br$	10.9	-4.5
	$-CH_2I$	14.2	-4.0
	$-CH_2OH$	14.2	-8.4
	$-CH_2OCH_2CH_3$	12.3	-8.8
	$-CH=CH_2$	13.6	-7.0
	$-phenyl$	12.5	-11.0
HAL	$-F$	24.9	-34.3
	$-Cl$	2.6	-6.1
	$-Br$	-7.9	-1.4
	$-I$	-38.1	7.0
O	$-OCH_3$	29.4	-38.9
	$-OCH_2CH_3$	28.5	-39.8
	$-OCH_2CH_2CH_2CH_3$	28.1	-40.4
	$-OCOCH_3$	18.4	-26.7
N	$-N^+(CH_3)_3$	19.8	-10.6
	$-N$-pyrrolidonyl	6.5	-29.2
	$-NO_2$	22.3	-0.9
	$-NC$	-3.9	-2.7
S	$-SCH_2$-phenyl	18.5	-16.4
	$-SO_2CH=CH_2$	14.3	7.9
O‖C⁄	$-CHO$	13.1	12.7
	$-COCH_3$	15.0	5.8
	$-COOH$	4.2	8.9
	$-COOCH_2CH_3$	6.3	7.0
	$-COCl$	8.1	14.0
	$-CN$	-15.1	14.2
	$-Si(CH_3)_3$	16.9	6.7
	$-SiCl_3$	8.7	16.1

The values listed on p. C90 can also be used to estimate the ^{13}C-chemical shifts of C=C double bonds with more than one substituent.

<u>Example</u>: Estimation of the ^{13}C-chemical shifts in 1-bromo-1-propene:

$$Br-\underset{a}{CH}=\underset{b}{CH}-CH_3$$

(a)			(b)		
base value:	123.3		base value:	123.3	
Z_1 (Br)	-7.9		Z_2 (Br)	-1.4	
Z_2 (CH$_3$)	-7.9		Z_1 (CH$_3$)	10.6	
estimated:	107.5		estimated:	132.5	
determined:	108.9	(cis)	determined:	129.4	(cis)
	104.7	(trans)		132.7	(trans)

¹³C-NMR

^{13}C-Chemical Shifts in Unsaturated Alicyclics

(δ in ppm relative to TMS)

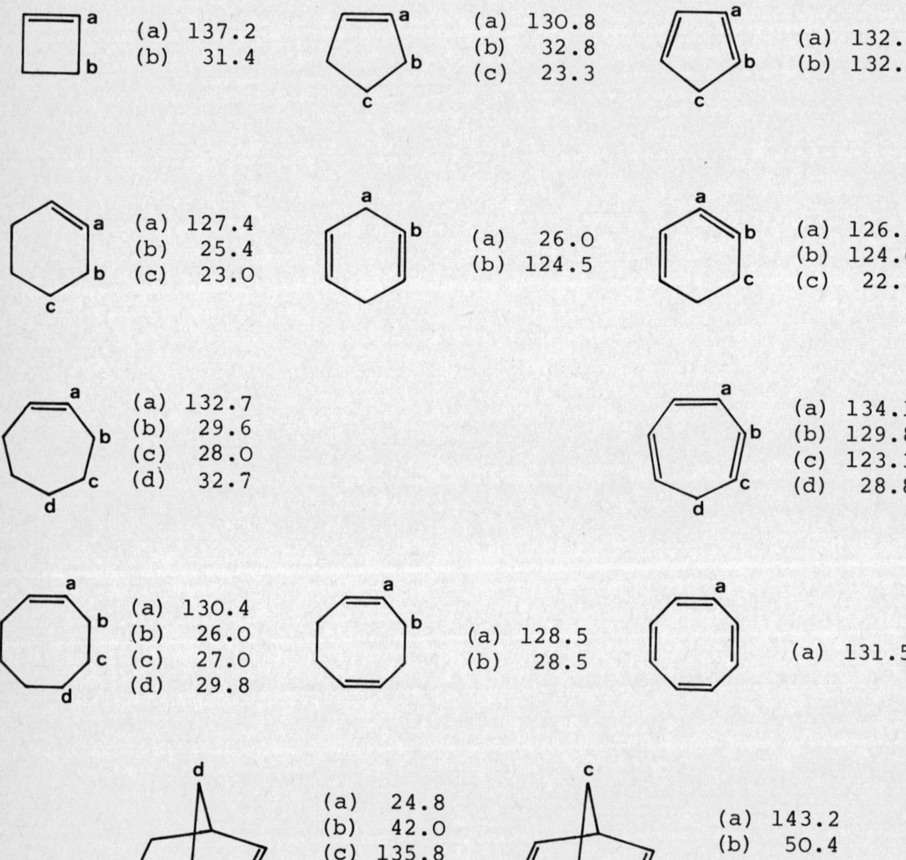

Monocyclic saturated alicyclics: see p. C50.
Condensed saturated alicyclics: see p. C75.

(a) 25.3	(a) 23.6
(b) 32.8	(b) 29.5
(c) 143.9	(c) 136.8
(d) 125.9	(d) 125.5
(e) 124.2	(e) 129.0

^{13}C-Chemical Shifts in Aliphatic Dienes (δ in ppm relative to TMS)

(a) allenes

$$\overset{b}{CH_2}=C=\overset{a}{CH_2}$$

(a) 74.8
(b) 213.5

(b) conjugated dienes

(a) 136.9
(b) 116.3

^{13}C-Chemical Shifts in Some Alkynes (δ in ppm relative to TMS)

$$H-C\equiv C-X$$
$$ab$$

X	a	b
-H	71.9	71.9
$-CH_3$	66.9	79.2
$-CH_2CH_2CH_2CH_3$	66.0	83.0
$-CH_2OH$	73.8	83.0
-phenyl	78.3	84.6
$-OCH_2CH_3$	23.2	89.4
$-SCH_2CH_2$	81.4	72.6

For the effect of a triple bond on the ^{13}C-chemical shift of neighboring carbon atoms see p. C10.

¹³C-Chemical Shifts in Some Aromatic Hydrocarbons

(δ in ppm relative to TMS)

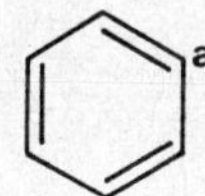

(a) 128.5

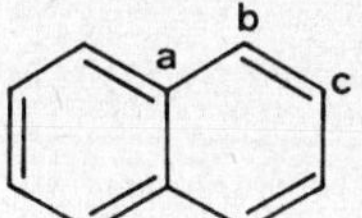

(a) 133.7
(b) 128.0
(c) 126.0

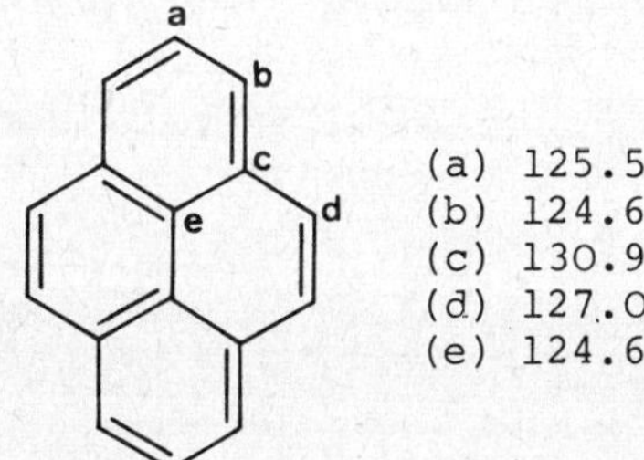

(a) 126.2
(b) 131.8
(c) 128.1
(d) 125.3

(a) 131.9
(b) 122.4
(c) 126.3
(d) 126.3
(e) 128.3
(f) 130.1
(g) 126.6

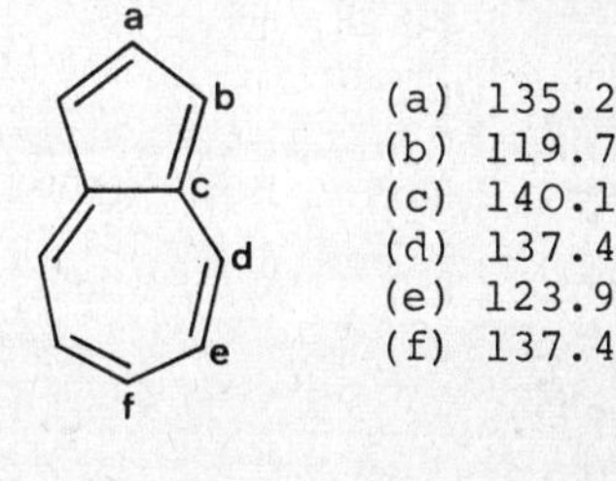

(a) 125.5
(b) 124.6
(c) 130.9
(d) 127.0
(e) 124.6

(a) 135.2
(b) 119.7
(c) 140.1
(d) 137.4
(e) 123.9
(f) 137.4

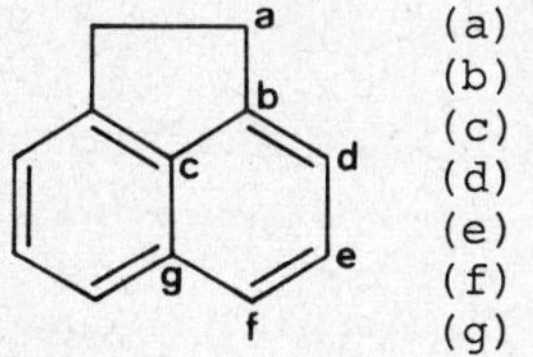

(a) 30.3
(b) 145.9
(c) 139.7
(d) 119.5
(e) 128.2
(f) 122.7
(g) 132.1

(a) 129.7
(b) 140.0
(c) 128.7
(d) 124.3
(e) 127.9
(f) 127.4
(g) 128.4

^{13}C-NMR

MONOSUBSTITUTED BENZENES

Effect of Substituents on the ^{13}C-Chemical Shifts in Monosubstituted Benzenes (δ in ppm relative to TMS)

$$\delta_{C_i} = 128.5 + z_i$$

	Substituent X	z_1	z_2	z_3	z_4
	$-H$	0.0	0.0	0.0	0.0
	$-CH_3$	9.3	0.6	0.0	-3.1
	$-CH_2CH_3$	15.7	-0.6	-0.1	-2.8
	$-CH(CH_3)_2$	20.1	-2.0	0.0	-2.5
	$-CH_2CH_2CH_2CH_3$	14.2	-0.2	-0.2	-2.8
	$-C(CH_3)_3$	22.1	-3.4	-0.4	-3.1
	$-\triangledown$	15.1	-3.3	-0.6	-3.6
	$-CH_2Cl$	9.1	0.0	0.2	-0.2
	$-CH_2Br$	9.2	0.1	0.4	-0.3
C	$-CF_3$	2.6	-3.1	0.4	3.4
	$-CH_2OH$	13.0	-1.4	0.0	-1.2
	$-\triangledown_O$	9.2	-3.1	-0.1	-0.5
	$-CH_2NH_2$	14.9	-1.6	-0.2	-2.0
	$-CH_2CN$	1.6	-0.7	0.5	-0.7
	$-CH=CH_2$	7.6	-1.8	-1.8	-3.5
	$-C\equiv CH$	-6.1	3.8	0.4	-0.2
	$-phenyl$	13.0	-1.1	0.5	-1.0
	$-F$	35.1	-14.3	0.9	-4.4
H	$-Cl$	6.4	0.2	1.0	-2.0
A	$-Br$	-5.4	3.3	2.2	-1.0
L	$-I$	-32.3	9.9	2.6	-0.4
	$-OH$	26.9	-12.7	1.4	-7.3
	$-O^-$	39.6	-8.2	1.9	-13.6
O	$-OCH_3$	30.2	-14.7	0.9	-8.1
	$-Ophenyl$	29.1	-9.5	0.3	-5.3
	$-OCOCH_3$	23.0	-6.4	1.3	-2.3
	$-NH_2$	19.2	-12.4	1.3	-9.5
	$-NHCH_3$	21.7	-16.2	0.7	-11.8
	$-N(CH_3)_2$	22.4	-15.7	0.8	-11.8
N	$-N(CH_2CH_3)_2$	19.3	-16.5	0.6	-13.0
	$-N(phenyl)_2$	19.3	-4.4	0.6	-5.9
	$-NHCOCH_3$	11.1	-9.9	0.2	-5.6
	$-NHNH_2$	22.8	-16.5	0.5	-9.6

Substituent X	z_1	z_2	z_3	z_4
N -N=N-phenyl	24.0	-5.8	0.3	2.2
-N$^+$≡N	-12.7	6.0	5.7	16.0
-NC	-1.8	-2.2	1.4	0.9
-NCO	5.7	-3.6	1.2	-2.8
-NO	37.4	-7.7	0.8	7.0
-NO$_2$	19.6	-5.3	0.8	6.0
S -SH	2.2	0.7	0.4	-3.1
-SCH$_3$	9.9	-2.0	0.1	-3.7
-SC(CH$_3$)$_3$	4.5	9.0	-0.3	0.0
-SO$_2$Cl	15.6	-1.7	1.2	6.8
-SO$_3$H	15.0	-2.2	1.3	3.8
-CHO	9.0	1.2	1.2	6.0
-COCH$_3$	9.3	0.2	0.2	4.2
-COOH	2.4	1.6	-0.1	4.8
-COO$^-$	7.6	0.8	0.0	2.8
-COOCH$_3$	2.1	1.2	0.0	4.4
-CONH$_2$	5.4	-0.3	-0.9	5.0
-COCl	4.6	2.9	0.6	7.0
-CN	-16.0	3.5	0.7	4.3
-P(phenyl)$_2$ [1]	8.7	5.1	-0.1	0.0
-Si(CH$_3$)$_3$	13.4	4.4	-1.1	-1.1

1) For ^{13}C-^{31}P-couplings see p. C245.

MULTIPLY SUBSTITUTED BENZENES

The ^{13}C-chemical shifts in multiply substituted benzene derivatives can be estimated using the ^{13}C-chemical shifts listed on p. C120 and C125 for monosubstituted benzenes.

Example: Estimation of the ^{13}C-chemical shifts in 3,5-dimethylnitrobenzene:

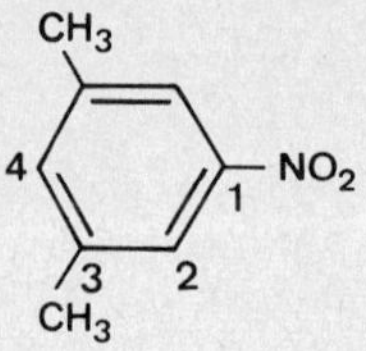

(C-1)		
	base value:	128.5
	$Z_1(NO_2)$	19.6
	$2Z_3(CH_3)$	0.0
	estimated:	148.1
	determined:	148.5

(C-2)		
	base value:	128.5
	$Z_2(NO_2)$	-5.3
	$Z_2(CH_3)$	0.6
	$Z_4(CH_3)$	-3.1
	estimated:	120.7
	determined:	121.7

(C-3)		
	base value:	128.5
	$Z_1(CH_3)$	9.3
	$Z_3(CH_3)$	0.0
	$Z_3(NO_2)$	0.8
	estimated:	138.6
	determined:	139.6

(C-4)		
	base value:	128.5
	$2Z_2(CH_3)$	1.2
	$Z_4(NO_2)$	6.0
	estimated:	135.7
	determined:	136.2

Larger discrepancies between estimated and experimentally determined values are to be expected if the substituents are ortho to each other.

^{13}C-Chemical Shifts in Five-Membered Heteroaromatic Rings

(δ in ppm relative to TMS)

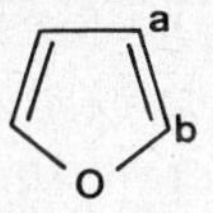

Furan (O)
(a) 109.9
(b) 143.0

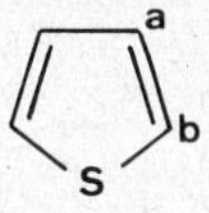

Thiophene (S)
(a) 126.4
(b) 124.9

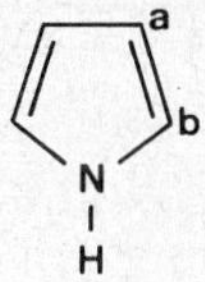

Pyrrole (N–H)
(a) 107.7
(b) 118.0

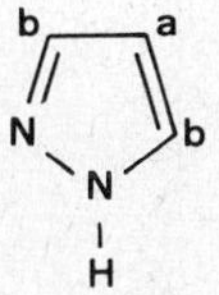

Pyrazole (N–N–H)
(a) 104.7
(b) 133.3

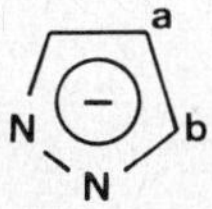

Pyrazolide anion (–)
(a) 103.4
(b) 138.5

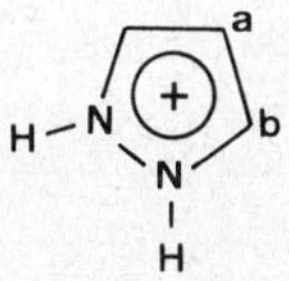

Pyrazolium cation (+)
(a) 109.0
(b) 135.0

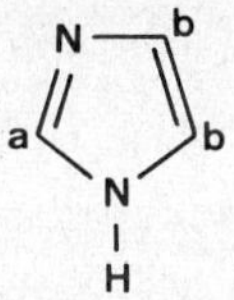

Imidazole (N–H)
(a) 136.2
(b) 122.3

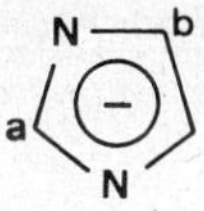

Imidazolide anion (–)
(a) 145.1
(b) 126.8

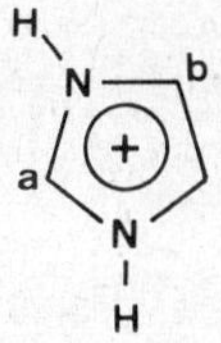

Imidazolium cation (+)
(a) 134.6
(b) 120.1

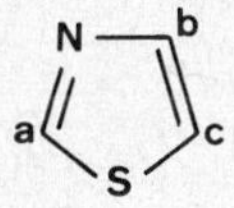

Thiazole (S)
(a) 152.7
(b) 143.2
(c) 118.6

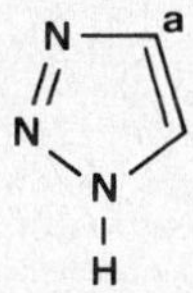

1,2,3-Triazole
(a) 130.4

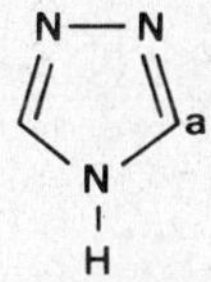

1,2,4-Triazole
(a) 126.8

Six-membered heteroaromatic rings: see p. C155.

^{13}C-NMR

MONOSUBSTITUTED PYRIDINES

<u>Effect of Substituents on the ^{13}C-Chemical Shifts in Monosubstituted Pyridines</u> (δ in ppm relative to TMS)

$$\delta_{C-2} = 149.8 + z_{i2}$$

$$\delta_{C-3} = 123.6 + z_{i3}$$

$$\delta_{C-4} = 135.7 + z_{i4}$$

$$\delta_{C-5} = 123.6 + z_{i5}$$

$$\delta_{C-6} = 149.8 + z_{i6}$$

2- or 6-Substituent (i = 2 or 6)	$z_{22}=z_{66}$	$z_{23}=z_{65}$	$z_{24}=z_{64}$	$z_{25}=z_{63}$	$z_{26}=z_{62}$
$-CH_3$	8.8	-0.6	0.2	-3.0	-0.4
$-CH_2CH_3$	13.6	-1.8	0.4	-2.9	-0.7
$-F$	14.4	-13.1	6.1	-1.5	-1.5
$-Cl$	2.3	0.7	3.3	-1.2	0.6
$-Br$	-6.7	4.8	3.3	-0.5	1.4
$-OH$	15.5	-3.5	-0.9	-16.9	-8.2
$-OCH_3$	15.3	-7.5	2.1	-13.1	-2.2
$-NH_2$	11.3	-14.7	2.3	-10.6	-0.9
$-NO_2$	8.0	-5.1	5.5	6.6	0.4
$-CHO$	3.5	-2.6	1.3	4.1	0.7
$-COCH_3$	4.3	-2.8	0.7	3.0	-0.2
$-CN$	-15.9	5.0	1.6	3.6	1.4

3- or 5-Substituent (i = 3 or 5)	$z_{32}=z_{56}$	$z_{33}=z_{55}$	$z_{34}=z_{54}$	$z_{35}=z_{53}$	$z_{36}=z_{52}$
$-CH_3$	1.3	9.0	0.2	-0.8	-2.3
$-CH_2CH_3$	-0.4	15.5	-0.6	-0.4	-2.7
$-F$	-11.5	36.2	-13.0	0.9	-3.9
$-Cl$	-0.3	8.2	-0.2	0.7	-1.4
$-Br$	2.1	-2.6	2.9	1.2	-0.9
$-I$	7.1	-28.4	9.1	2.4	0.3
$-OH$	-10.7	31.4	-12.2	1.3	-8.6
$-NH_2$	-11.9	21.5	-14.2	0.9	-10.8
$-CHO$	2.4	7.9	0.0	0.6	5.4
$-COCH_3$	3.5	8.6	-0.5	-0.1	0.0
$-CONH_2$	2.7	6.0	1.3	1.3	-1.5
$-CN$	3.6	-13.7	4.4	0.6	4.2

4-Substituent (i = 4)	$z_{42}=z_{46}$	$z_{43}=z_{45}$	z_{44}
$-CH_3$	0.5	0.8	10.8
$-CH_2CH_3$	-0.1	-0.4	17.0
$-CH(CH_3)_2$	0.4	-1.8	21.4
$-C(CH_3)_3$	0.1	-3.4	23.4
$-CH=CH_2$	0.3	-2.9	8.6
$-F$	2.7	-11.8	33.0
$-Br$	3.0	3.4	-3.0
$-NH_2$	0.9	-13.8	19.6
$-CHO$	1.7	-0.6	5.5
$-COCH_3$	1.6	-2.6	6.8
$-CN$	2.1	2.2	-15.7

MULTIPLY SUBSTITUTED PYRIDINES

The ^{13}C-chemical shifts in multiply substituted pyridines can be estimated using the ^{13}C-chemical shifts listed on p. C140 and C145 for monosubstituted pyridines.

<u>Example</u>: Estimation of the ^{13}C-chemical shifts in 2,5-dimethyl-pyridine:

(C-2) base value:	149.8		(C-3) base value:	123.6
$Z_{22}(CH_3)$	8.8		$Z_{23}(CH_3)$	-0.6
$Z_{52}(CH_3)$	-2.3		$Z_{53}(CH_3)$	-0.8
estimated:	156.3		estimated:	122.2
determined:	155.2		determined:	122.5

(C-4) base value:	135.7		(C-5) base value:	123.6
$Z_{24}(CH_3)$	0.2		$Z_{55}(CH_3)$	9.0
$Z_{54}(CH_3)$	0.2		$Z_{25}(CH_3)$	-3.0
estimated:	136.1		estimated:	129.6
determined:	136.7		determined:	129.6

(C-6) base value:	149.8
$Z_{56}(CH_3)$	1.3
$Z_{26}(CH_3)$	-0.4
estimated:	150.7
determined:	149.4

^{13}C-Chemical Shifts in Six-Membered Heteroaromatic Rings

(δ in ppm relative to TMS)

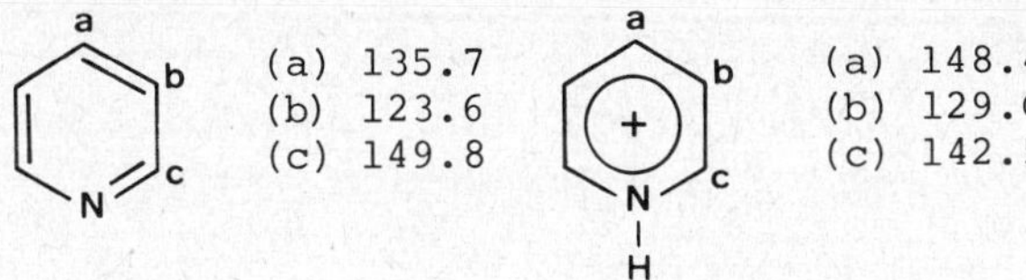

Substituted pyridines: see p. C140.

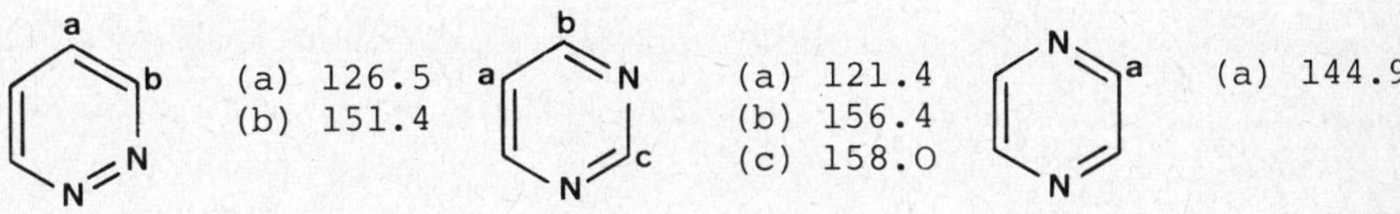

^{13}C-NMR

CONDENSED HETEROAROMATICS

^{13}C-Chemical Shifts in Condensed Heteroaromatics

(δ in ppm relative to TMS)

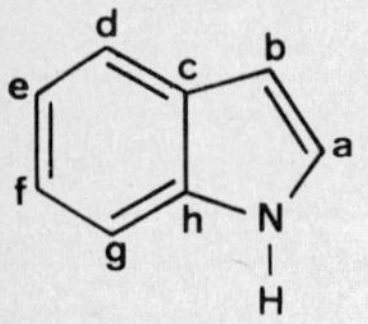

(a) 124.1	(e) 121.7
(b) 102.1	(f) 119.6
(c) 127.6	(g) 111.0
(d) 120.5	(h) 135.5

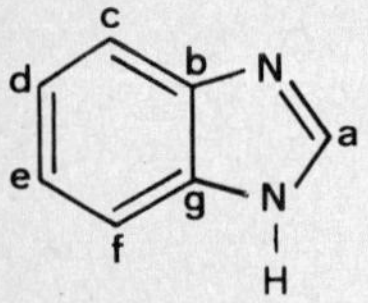

(a)	141.5
(b,g)	137.9
(c,f)	115.4
(d,e)	122.9

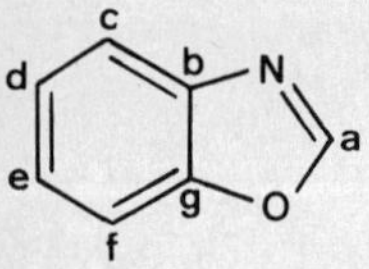

(a) 152.6	(e) 124.4
(b) 140.1	(f) 110.8
(c) 120.5	(g) 150.0
(d) 125.4	

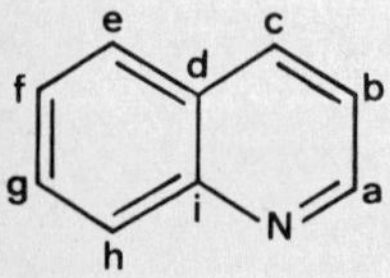

(a) 150.0	(d) 128.0	(g) 129.2
(b) 120.8	(e) 127.6	(h) 129.2
(c) 135.7	(f) 126.3	(i) 148.1

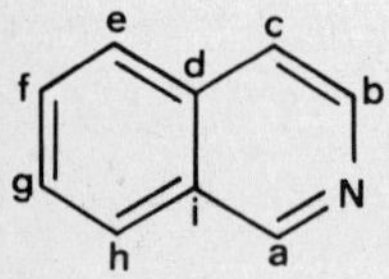

(a) 152.2	(d) 135.5	(g) 127.0
(b) 142.7	(e) 126.2	(h) 127.3
(c) 120.2	(f) 130.1	(i) 128.5

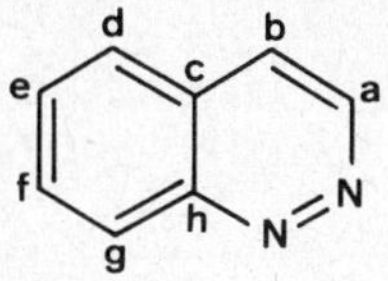

(a) 146.1   (d) 128.0   (g) 129.5
(b) 124.7   (e) 132.2   (h) 151.0
(c) 126.9   (f) 132.1

(a) 155.9   (d) 127.4   (g) 128.6
(b) 160.7   (e) 127.9   (h) 150.1
(c) 125.2   (f) 134.1

(a) 144.8   (c) 129.6
(b) 142.8   (d) 129.4

(a) 152.0   (c) 126.7
(b) 126.7   (d) 133.1

(a) 149.1   (d) 128.3   (g) 135.8
(b) 130.3   (e) 129.5
(c) 125.5   (f) 126.6

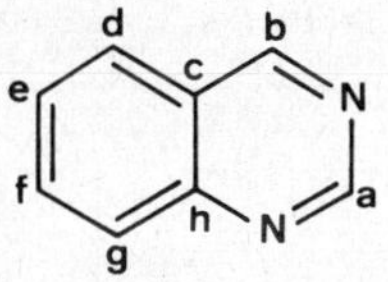

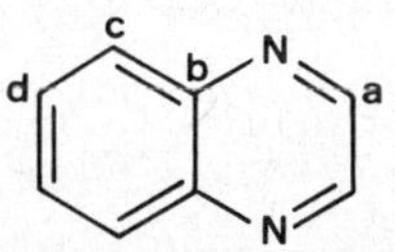

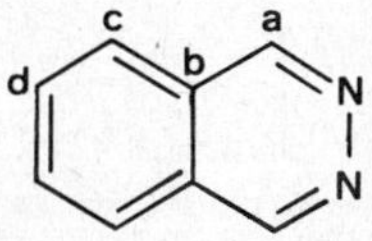

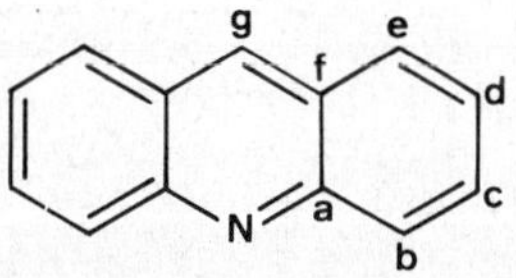

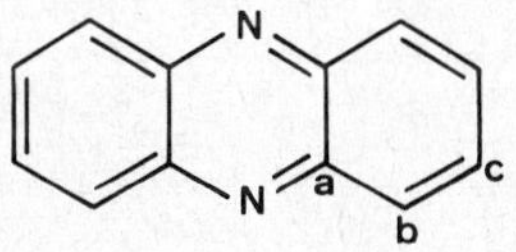

(a) 144.0
(b) 130.9
(c) 130.2

(a) 152.0   (d) 128.4
(b) 154.9   (e) 144.8
(c) 147.9

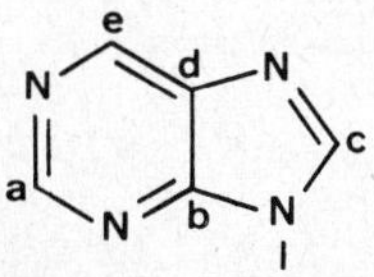

^{13}C-Chemical Shifts for Carbonyl Groups (δ in ppm relative to TMS)

R	R-CHO	R-COCH$_3$	R-COOH	R-COO$^-$	R-COOCH$_3$	R-CONH$_2$	R-COOCOR	R-COCl
-H		199.7	166.3	171.3	161.6	165.5		
-CH$_3$	199.7	206.0	178.1	181.7	170.7	172.7	167.3	168.6
-CH$_2$CH$_3$	201.8	207.6	180.4	185.1	173.3	177.2	170.3	174.7
-CH(CH$_3$)$_2$	204.0	211.8	184.1		175.7		172.8	178.0
-C(CH$_3$)$_3$		213.5	185.9	188.6	178.9	180.9	173.9	180.3
-n-C$_8$H$_{17}$	202.6	207.9	180.7	183.1	174.4	176.3	169.4	173.8
-CH$_2$Cl	193.3	200.7	173.8	174.7	167.8	168.3	162.1	167.7
-CHCl$_2$		193.6	170.3	170.6	165.1		157.6	165.5
-CCl$_3$	175.9	186.3	167.1		161.0		154.1	
-cyclohexyl	201.8	209.4	182.1	185.4	175.3			176.3
-CH=CH$_2$	192.4	197.2	171.7	179.3	165.5	168.3		165.6
-phenyl	192.0	197.6	172.6	175.6	166.8	169.7	162.8	168.0

^{13}C-Chemical Shifts in Cyclic Ketones, Lactones and Lactams

(δ in ppm relative to TMS)

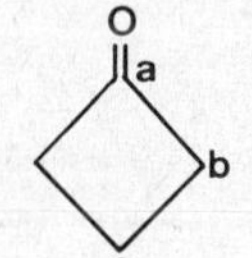

(a) 208.2
(b) 46.9
(c)   9.0

(a) 171.2

(a) 213.9
(b) 37.0
(c) 22.3

(a) 177.9
(b) 27.7
(c) 22.2
(d) 68.6

(a) 179.4
(b) 30.3
(c) 20.8
(d) 42.4

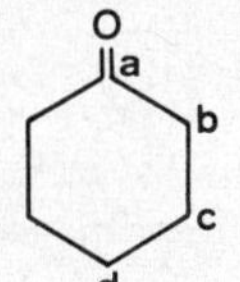

(a) 208.8
(b) 40.7
(c) 26.8
(d) 24.1

(a) 175.2
(b)
(c)
(d)
(e)

(a) 173.0
(b) 31.4
(c) 20.8*)
(d) 22.2*)
(e) 42.0

(a) 211.7
(b) 42.7
(c) 29.7
(d) 23.5

(a) 176.2
(b) 34.6
(c) 23.0
(d) 28.9*)
(e) 29.4*)
(f) 69.2

(a) 169.4
(b) 32.3
(c) 21.6*)
(d) 23.3*)
(e) 49.9
(f) 34.4

(a) 209.0

(a) 198.0

(a) 179.4
(b) 36.9
(c) 23.5
(d) 30.0*)
(e) 30.8*)
(f) 42.6

*) assignment uncertain

^{13}C-NMR

DICARBONYL COMPOUNDS

^{13}C-Chemical Shifts in Quinones (δ in ppm relative to TMS)

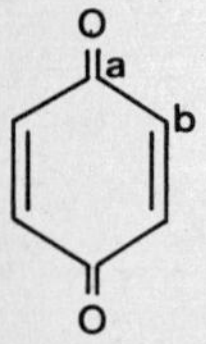

(a) 187.0
(b) 136.4

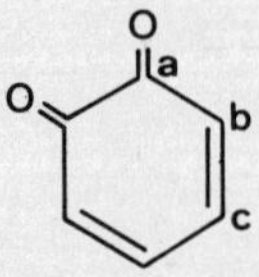

(a) 180.4
(b) 130.8
(c) 139.7

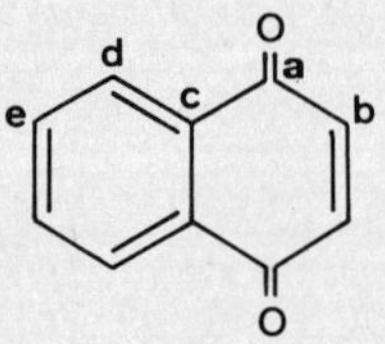

(a) 184.7
(b) 138.5
(c) 131.8
(d) 126.2
(e) 133.7

^{13}C-Chemical Shifts in Dicarboxylic Acids and Cyclic Derivatives

(δ in ppm relative to TMS)

COOH — 160.9
COOH

COOH — 170.4
CH$_2$ — 41.4
COOH

COOH — 176.4
CH$_2$ — 30.0
CH$_2$
COOH

COOH — 177.1
CH$_2$ — 33.7
CH$_2$ — 22.0
CH$_2$
COOH

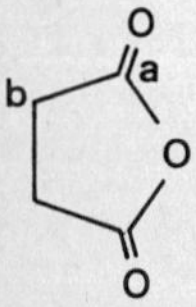

(a) 172.9
(b) 28.6

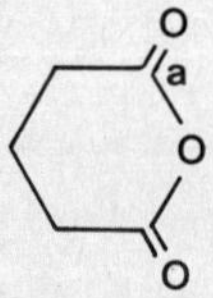

(a) 168.2

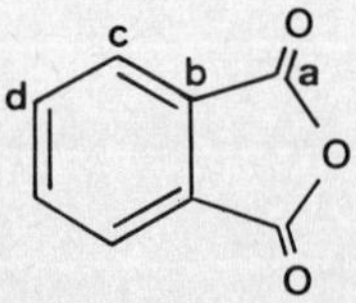

(a) 163.1
(b) 131.1
(c) 125.3
(d) 136.1

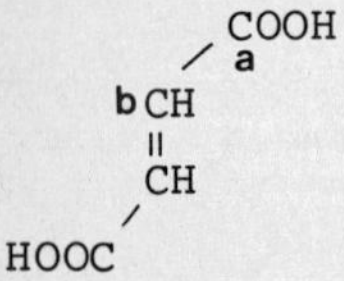

(a) 166.9
(b) 134.5

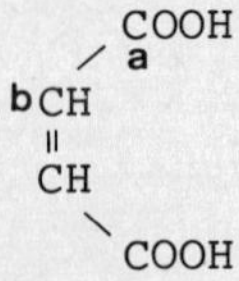

(a) 167.4
(b) 130.8

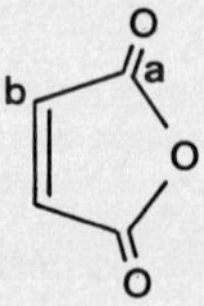

(a) 164.3
(b) 136.6

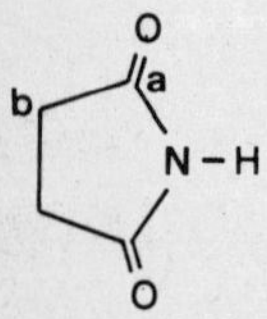

(a) 183.6
(b) 30.3

^{13}C-Chemical Shifts in Carbonic Acid Derivatives

(δ in ppm relative to TMS)

CH_3 | O | CO 156.5 | O | CH_3	CH_3 13.6 | CH_2 19.1 | CH_2 30.9 | CH_2 67.6 | O | CO 155.9 | OC_4H_9	CH_3 14.7 | CH_2 60.7 | O | CO 157.8 | NH | CH_3 27.4

CO_3^{--} 168.5

NH_2
\|
CO 161.2
\|
NH_2

$(CH_3)_2 N{-}CO{-}N(CH_3)_2$ CH_2: 38.5, CO: 165.4

$(CH_3)_2 N{-}CS{-}N(CH_3)_2$ CH_3: 43.0, CS: 193.9

ethylene carbonate (a) 156.7

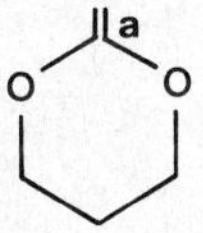

1,3-dioxan-2-one (a) 155.2

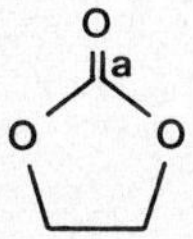

^{13}C-NMR

NITRILES, ISONITRILES

^{13}C-Chemical Shifts in Nitriles

(δ in ppm relative to TMS)

The ^{13}C-Chemical shift of the nitrile carbon is found between 115 and 125 ppm.

CN	117.7	CN	120.8	CN	123.7	CN	125.1
CH$_3$	1.3	CH$_2$	10.6	CH	22.4	C	28.3
		CH$_3$	10.8	(CH$_3$)$_2$	27.4	(CH$_3$)$_3$	36.5

CN	117.2	a CN (b, c, d, e ring)	(a) 122.4 (b) 28.3 (c) 30.1 (d) 24.6 (e) 25.8	a CN (b, c, d, e ring)	(a) 118.7 (b) 112.5 (c) 132.0 (d) 129.2 (e) 132.8
CH	107.8				
CH$_2$	137.5				

^{13}C-Chemical Shifts and ^{13}C-^{14}N-Couplings in Isonitriles

(δ in ppm relative to TMS)

Because of the symmetrical electron distribution around the nitrogen atom the ^{13}C-^{14}N-coupling can be observed in the ^{13}C-NMR spectra of isonitriles.

C$^-$≡N$^+$	158.2 (J_{CN} = 5.8 Hz)	C$^-$≡N$^+$	156.8 (J_{CN} = 5.3 Hz)
CH$_3$	26.8 (J_{CN} = 7.5 Hz)	CH$_2$	36.4 (J_{CN} = 6.5 Hz)
		CH$_3$	15.3 (J_{CN} = ∿0 Hz)

C$^-$≡N$^+$	165.7 (J_{CN} = 5.0 Hz)	a C$^-$≡N$^+$ (b, c, d, e ring)	(a) 165.7 (J_{CN} = 5.2 Hz) (b) 126.7 (J_{CN} = 13.2 Hz) (c) 126.3 (d) 129.9 (e) 129.4 } (J_{CN} = 0 Hz)
CH	119.4 (J_{CN} = 11.7 Hz)		
CH$_2$	120.6 (J_{CN} = ∿0 Hz)		

C190

^{13}C-Chemical Shifts in Imines (δ in ppm relative to TMS)

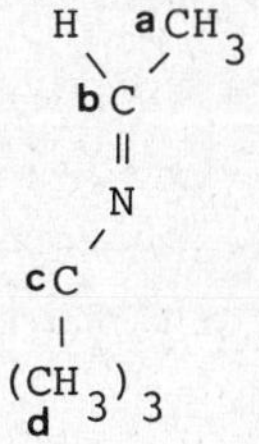

(a) 22.6
(b) 154.2
(c) 56.6
(d) 29.7

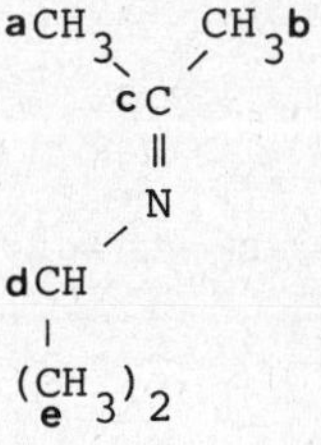

(a) 17.8
(b) 29.3
(c) 163.4
(d) 50.6
(e) 23.6

^{13}C-Chemical Shifts in Oximes (δ in ppm relative to TMS)

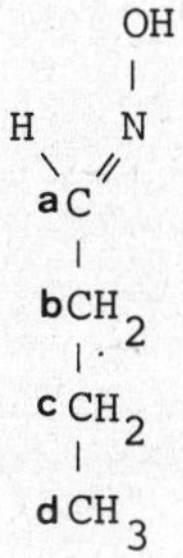

(a) 152.3
(b) 31.5
(c) 20.2
(d) 13.9

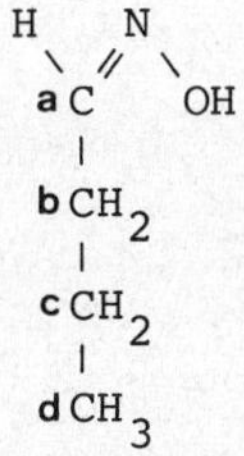

(a) 151.9
(b) 27.1
(c) 19.6
(d) 13.6

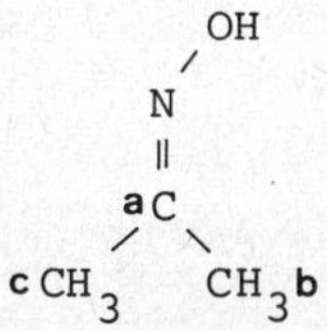

(a) 155.5
(b) 20.9
(c) 14.3

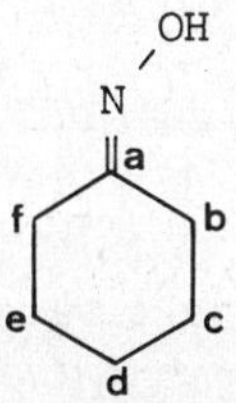

(a) 159.4
(b) 27.5
(c) 26.1
(d) 24.6
(e) 26.3
(f) 32.3

^{13}C-Chemical Shifts in Isocyanates (δ in ppm relative to TMS)

NCO	121.5
CH$_3$	26.3

NCO	125.0 (broad)
CH$_2$	43.3
CH$_2$	34.2
CH$_2$	20.4
CH$_3$	13.6

¹³C-NMR

^{13}C-Chemical Shifts in Sulfur Containing Carbonyl Derivatives

(δ in ppm relative to TMS)

Thiocarbonyl groups exhibit a ^{13}C-chemical shift which is $\sim$30 ppm larger than that of the corresponding carbonyl groups. The carbonyl groups in thioacids and their esters are similarly more deshielded by about 20 ppm than is the case in carboxylic acids or esters.

NH_2		SH		CH_3	13.6
CS	205.6	CO	194.5	CH_2	22.2
CH_3	33.3	CH_3	32.6	CH_2	32.1
				CH_2	28.4
				S	
				CO	194.1
				CH_3	30.1

NH_2 — CS (a)   202.1

(a) 140.1
(b) 128.1
(c) 128.8
(d) 132.1

^{13}C-Chemical Shifts in Thiocyanates and Isothiocyanates

(δ in ppm relative to TMS)

SCN	111.8	SCN^-	133.3	NCS	128.7	NCS	131 (broad)
CH_2	28.7			CH_3	29.3	CH_2	45.0
CH_3	15.4					CH_2	32.3
						CH_2	20.0
						CH_3	13.3

^{13}C-Chemical Shifts in Amino Acids

(δ in ppm relative to TMS as external reference, solvent: water)

$NH_2-\underset{b}{CH_2}-\underset{a}{COOH}$

(a) 173.5
(b) 42.5

$NH_2-\underset{c}{CH_2}-\underset{b}{CH_2}-\underset{a}{COOH}$

(a) 183.4
(b) 35.7
(c) 38.5

$\underset{c}{CH_3}$
|
$NH_2-\underset{b}{CH}-\underset{a}{COOH}$

(a) 176.8
(b) 51.6
(c) 17.3

$\overset{d,e}{(CH_3)_2}$
|
$\underset{c}{CH}$
|
$NH_2-\underset{b}{CH}-\underset{a}{COOH}$

(a) 175.3
(b) 61.6
(c) 30.2
(d) 19.1
(e) 17.9

$\overset{e,f}{(CH_3)_2}$
|
$\underset{d}{CH}$
|
$\underset{c}{CH_2}$
|
$NH_2-\underset{b}{CH}-\underset{a}{COOH}$

(a) 176.6
(b) 54.7
(c) 41.0
(d) 25.4
(e) 23.2
(f) 22.1

$\underset{f}{CH_3}$
|
$\underset{e}{CH_2}$
| d
$\underset{c}{CH}-CH_3$
|
$NH_2-\underset{b}{CH}-\underset{a}{COOH}$

(a) 175.2
(b) 60.9
(c) 39.7
(d) 15.9
(e) 25.7
(f) 12.5

OH
|
$\underset{c}{CH_2}$
|
$NH_2-\underset{b}{CH}-\underset{a}{COOH}$

(a) 173.1
(b) 57.4
(c) 61.3

$\underset{d}{CH_3}$
|
$\underset{c}{CH}-OH$
|
$NH_2-\underset{b}{CH}-\underset{a}{COOH}$

(a) 174.0
(b) 61.5
(c) 67.1
(d) 20.5

SH
|
$\underset{c}{CH_2}$
|
$HCl \cdot NH_2-\underset{b}{CH}-\underset{a}{COOH}$

(a) 172.0
(b) 57.5
(c) 27.4

$S-\overset{e}{CH_3}$
|
$\underset{d}{CH_2}$
|
$\underset{c}{CH_2}$
|
$NH_2-\underset{b}{CH}-\underset{a}{COOH}$

(a) 175.3
(b) 55.3
(c) 31.0
(d) 30.1
(e) 15.2

^{13}C-NMR

AMINO ACIDS

```
HCl·NH2-CH-COOH
         |
         CH2
         |
         S
         |
         S
         |
        c CH2
         |
HCl·NH2-CH-COOH
        b  a
```

(a) 175.6
(b) 54.7
(c) 39.0

```
        g
      /   \
     |     | f
     |     |
      \   / e
        d
        |
       c CH2
        |
  NH2-CH-COOH
      b  a
```

(a) 175.0
(b) 57.3
(c) 37.5
(d) (~145)
(e) 131.1
(f) 130.7
(g) 129.5

```
        OH
        |
        g
      /   \
     |     | f
     |     |
      \   / e
        d
        |
       c CH2
        |
  NH2-CH-COOH
      b  a
```

(a) 175.0
(b) 57.3
(c) 37.5
(d) (~138)
(e) 130.5
(f) 117.5
(g) 156.3

```
 d COOH
   |
 c CH2
   |
NH2-CH-COOH
    b  a
```

(a) 175.5
(b) 53.2
(c) 37.6
(d) 178.8

```
 e COOH
   |
 d CH2
   |
 c CH2
   |
NH2-CH-COOH
    b  a
```

(a) 175.6
(b) 55.7
(c) 28.1
(d) 34.5
(e) 182.3

```
   NH2·HCl
    |
 e CH2
    |
 d CH2
    |
 c CH2
    |
NH2-CH-COOH
    b  a
```

(a) 179.4
(b) 56.6
(c) 29.2
(d) 24.5
(e) 41.1

```
   NH2
    |
 f CH2
    |
 e CH2
    |
 d CH2
    |
 c CH2
    |
NH2-CH-COOH
    b  a
```

(a) 175.4
(b) 55.3
(c) 27.2
(d) 22.4
(e) 30.7
(f) 40.0

C210

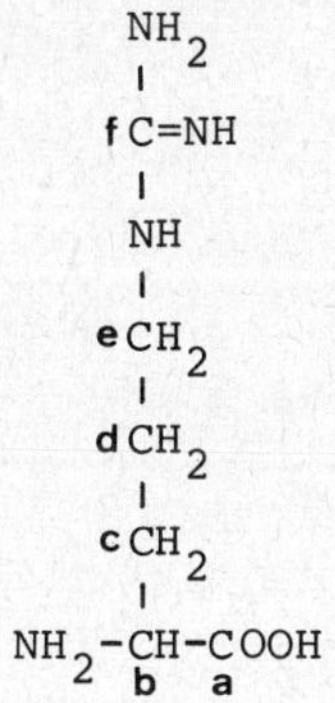

NH2
f C=NH
NH
e CH2
d CH2
c CH2
NH2-CH-COOH
b a

(a) 175.2
(b) 55.1
(c) 28.5
(d) 24.9
(e) 41.5
(f) 157.5

d
c
e
b
N
COOH
a
H

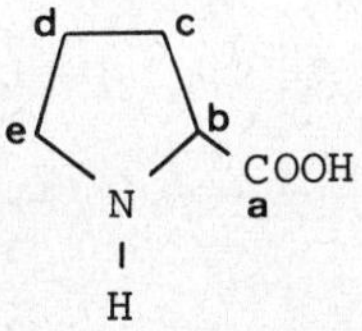

(a) 174.6
(b) 61.6
(c) 29.7
(d) 24.4
(e) 46.5

HO
d
c
e
b
N
COOH
a
H

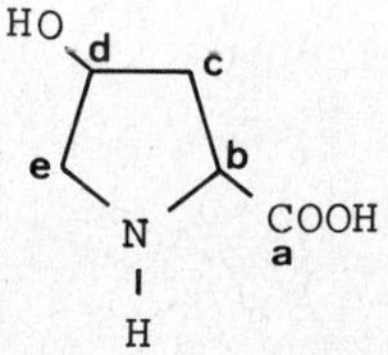

(a) 174.9
(b) 60.7
(c) 38.2
(d) 70.9
(e) 53.9

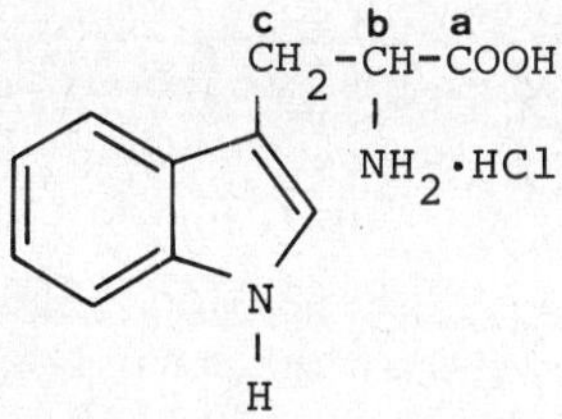

c b a
CH2-CH-COOH
NH2·HCl
N
H

(a) 174.7
(b) 56.0
(c) 28.2

N
e
f
d
c b a
CH2-CH-COOH
N
NH2
H

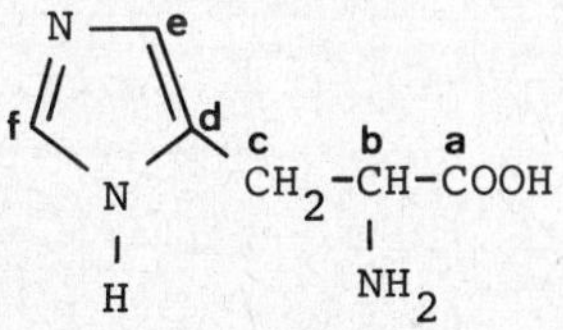

(a) 174.9
(b) 58.8
(c) 29.0
(d) (~135)
(e) 118.2
(f) 137.2

^{13}C-NMR

^{13}C-^{1}H-COUPLING CONSTANTS

<u>Coupling Across One Bond</u> ($^{1}J_{C-H}$ in Hz)

CH_4 125

CH_3Cl 151   △ 161 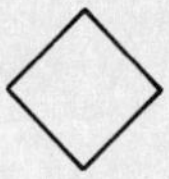   ◇ 134

CH_2=CH_2 156

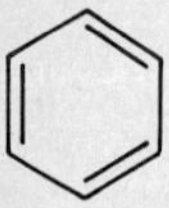

⬡ 159 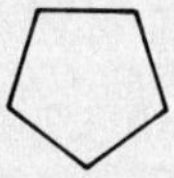   ⬠ 128 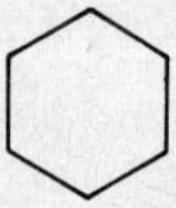   ⬡ 123

HC≡CH   249

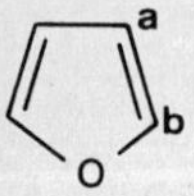

furan (a) 167 (b) 198

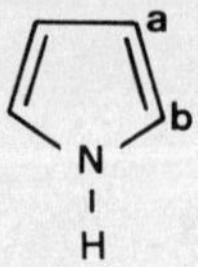

pyrrole (a) 170 (b) 182

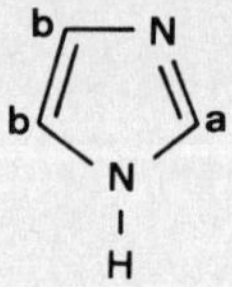

imidazole (a) 208 (b) 199

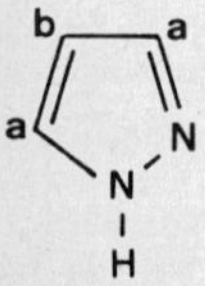

pyrazole (a) 190 (b) 178

1,2,3-triazole (a) 205

1,2,4-triazole (a) 208

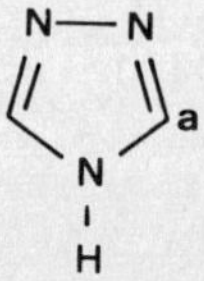

pyridine (a) 152 (b) 163 (c) 179

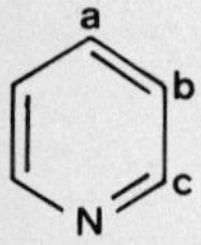

Additivity Rule for Estimating the ^{13}C-^{1}H-Coupling Constants Across a Bond in Aliphatic Compounds ($^{1}J_{C-H}$ in Hz)

$$J_{CH\ Z_1Z_2Z_3} = 125.0 + \sum_i Z_i$$

Substituent	Increments Z_i
-H	0.0
-CH$_3$	1.0
-C(CH$_3$)$_3$	-3.0
-CH$_2$Cl	3.0
-CH$_2$Br	3.0
-CH$_2$I	7.0
-CHCl$_2$	6.0
-CCl$_3$	9.0
-C≡CH	7.0
-phenyl	1.0
-F	24.0
-Cl	27.0
-Br	27.0
-I	26.0
-OH	18.0
-O-phenyl	18.0
-NH$_2$	8.0
-NHCH$_3$	7.0
-N(CH$_3$)$_2$	6.0
-SOCH$_3$	13.0
-CHO	2.0
-COCH$_3$	-1.0
-COOH	5.5
-CN	11.0

Example: Estimating the ^{13}C-^{1}H-coupling constant in CHCl$_3$

J = 125.0 + 3 · 27.0 = 206.0 Hz (determined: 209.0 Hz)

^{13}C-NMR

^{13}C-^{1}H-COUPLING CONSTANTS

Coupling Over Two Bonds ($|^{2}J_{C-H}|$ in Hz)

Typical ranges:

H-C-^{13}C-	1 - 6	
H-C=^{13}C-	0 - 16	
H-C≡^{13}C-	40 - 60	
H-C-^{13}CO-	5 - 8	
H-CO-^{13}C-	20 - 50	

Examples:

H-CH$_2$-^{13}CH$_3$	4.5
H-CH=^{13}CH$_2$	2.4
H-CCl=^{13}CHCl	16.0
H-C≡^{13}CH	49.3
H-CH$_2$-^{13}CO-CH$_3$	5.5
H-CO-^{13}CH$_3$	26.7

For the aromatic ring (H and ^{13}C in ortho position): 1 - 4 (example: 1.0)

R: any substituent

The ^{13}C-^{1}H-coupling constants over two bonds with an aldehyde or acetylene proton are so large that they can be observed as split peaks in "off resonance" decoupled spectra.

<u>Coupling Over Three Bonds</u> ($|^{3}J_{C-H}|$ in Hz)

$$|^{3}J_{C-H}| = 0\ldots\ldots 10$$

These coupling constants depend on the dihedral angle in the same way as the vicinal ^{1}H-^{1}H-coupling constants (see p. H20).

~ 3

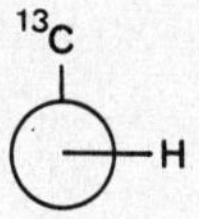

~ 0

~ 8

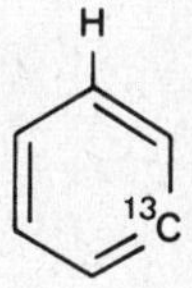

7 - 10

^{13}C-NMR

^{19}F–^{13}C-COUPLING CONSTANTS

<u>Coupling Between ^{13}C and ^{19}F</u> (spin quantum number I = 1/2, natural abundance 100 %) (|J| in Hz)

	|J|
F	
CH$_2$	164,8
CH$_2$	18,3
CH$_2$	6,2
CH$_2$	~ 0
C$_4$H$_9$	

	|J|
CF$_3$	294
COOH	46

	|J|
(a)	171
(b)	19
(c)	5
(d)	0

	|J|
(a)	245
(b)	21
(c)	8
(d)	3

	|J|
(a)	271.7
(b)	32.3
(c)	3.9
(d)	1.3
(e)	~ 0

	|J|
(a)	236.3
(b)	37.6
(c)	7.5
(d)	4.2
(e)	14.9

	|J|
(a)	22.6
(b)	255.1
(c)	17.7
(d)	3.7
(e)	3.7

	|J|
(a)	6.4
(b)	16.1
(c)	261.8

^{13}C-Chemical Shifts (δ in ppm relative to TMS) and ^{31}P-^{13}C-Coupling Constants ($|J|$ in Hz) in Some Phosphorous Compounds

$$X-CH_2CH_2CH_2CH_3$$

Substituent X	X-CH$_2$-		-CH$_2$-		-CH$_2$-		CH$_3$	
	δ	$|J|$	δ	$|J|$	δ	$|J|$	δ	$|J|$
-PCl$_2$	42.9	44	25.1	14	23.4	11	13.7	0
-P(n-C$_4$H$_9$)$_2$	28.6	14	27.9	15	24.8	10	14.0	0
-P$^+$(n-C$_4$H$_9$)$_3$	18.7		23.7		24.1		13.3	
-PO(n-C$_4$H$_9$)$_2$	27.8	66	24.0	5	24.4	13	13.6	0
-PS(n-C$_4$H$_9$)$_2$	30.9	51	24.6	4	24.0	16	13.6	0
-OP(O-n-C$_4$H$_9$)$_2$	61.9	11	33.4	5	19.1	0	13.7	0
-OPO(O-n-C$_4$H$_9$)$_2$	67.2	6	32.6	7	18.9	0	13.6	0

Substituent X	a		b		c		d	
	δ	$|J|$	δ	$|J|$	δ	$|J|$	δ	$|J|$
-P(phenyl)$_2$	137.2	12	133.6	20	128.4	7	128.5	0
-O-P(O-phenyl)$_2$	151.5	4	124.1	7	129.5	0	124.1	0
-O-PO(O-phenyl)$_2$	150.4	8	120.1	5	129.7	0	125.5	0

^{13}C-NMR

SOLVENTS

^{13}C-NMR Spectra of Common Solvents

(δ in ppm relative to TMS, idealized line intensities)

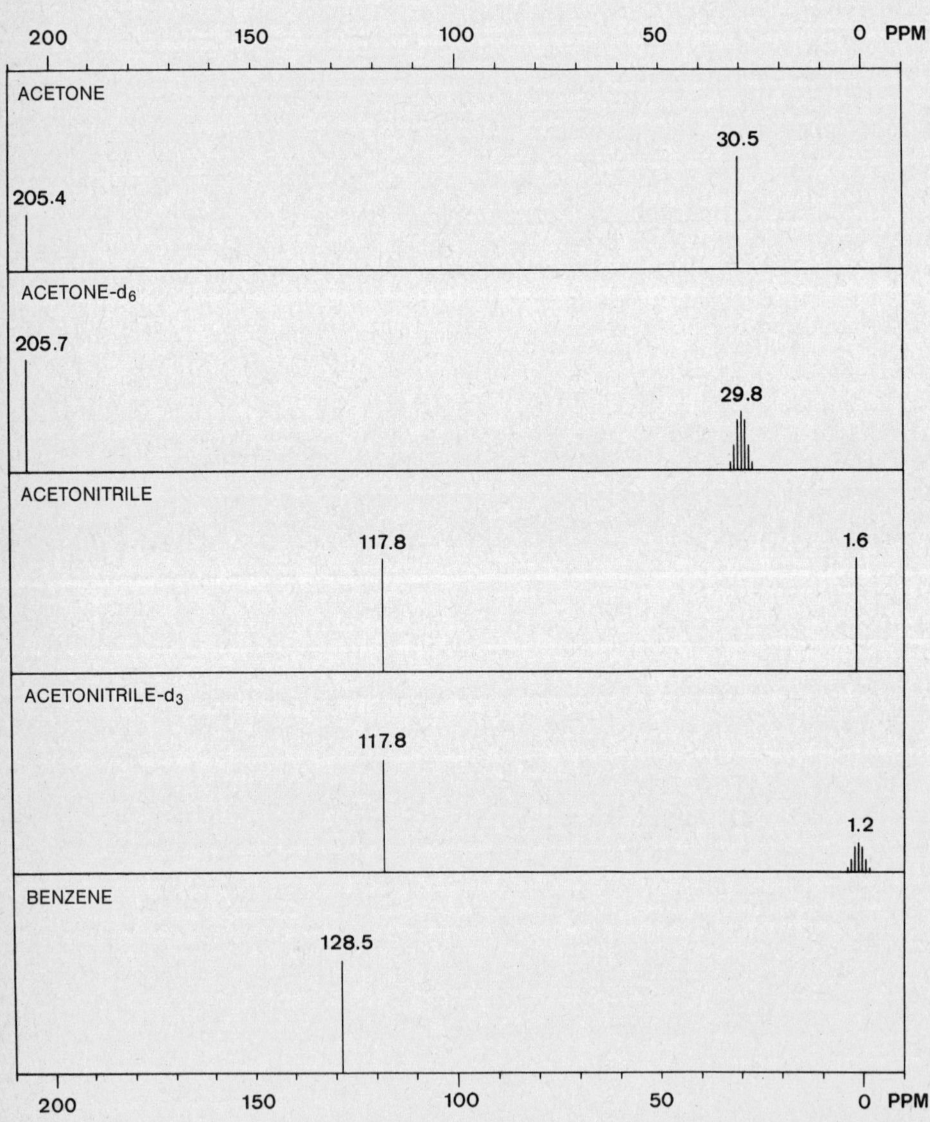

C250

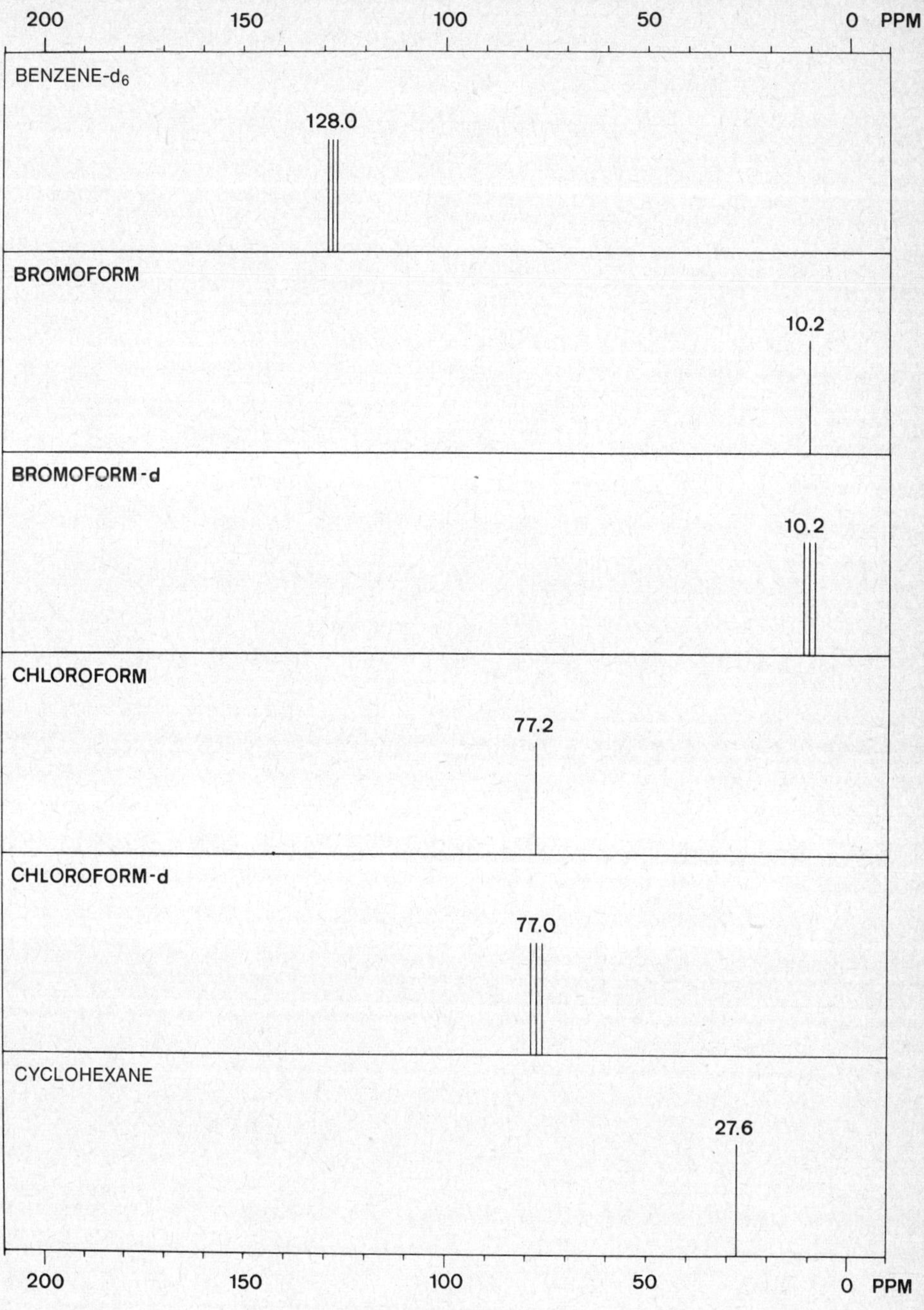
200 150 100 50 0 PPM
BENZENE-d6
128.0
BROMOFORM
10.2
BROMOFORM-d
10.2
CHLOROFORM
77.2
CHLOROFORM-d
77.0
CYCLOHEXANE
27.6
200 150 100 50 0 PPM

^{13}C-NMR

SOLVENTS

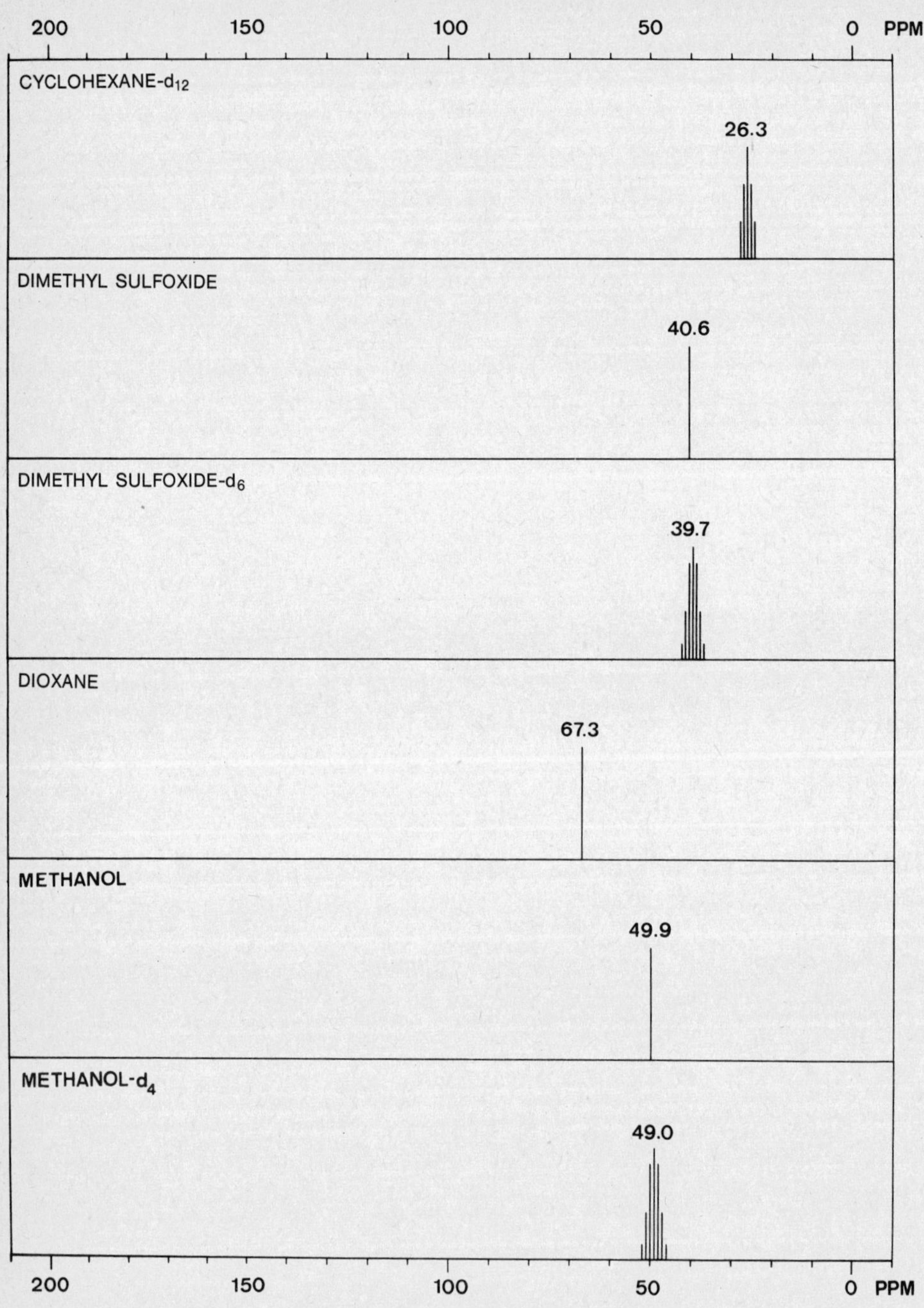

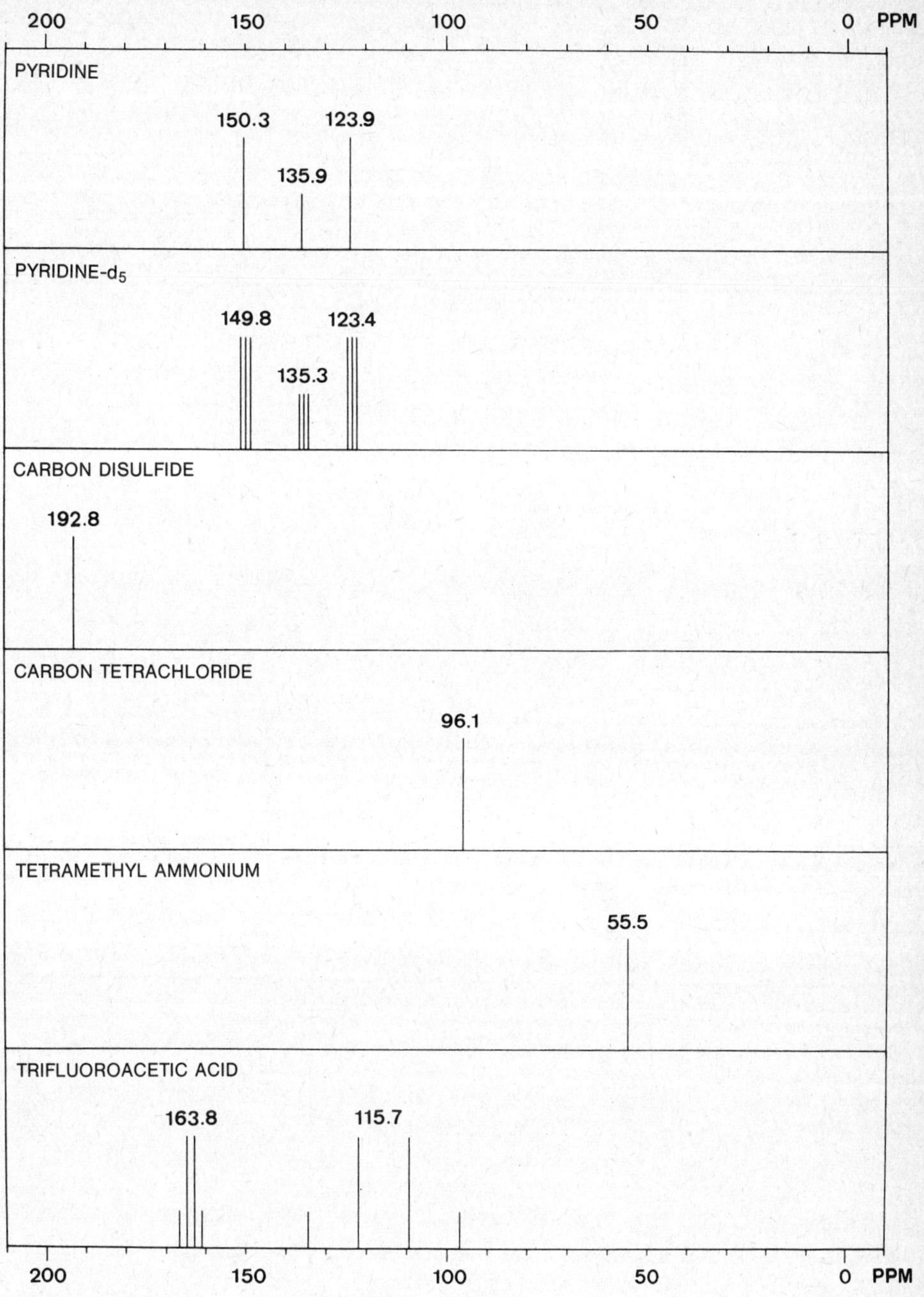
200   150   100   50   0 PPM
PYRIDINE
150.3   123.9
135.9
PYRIDINE-d5
149.8   123.4
135.3
CARBON DISULFIDE
192.8
CARBON TETRACHLORIDE
96.1
TETRAMETHYL AMMONIUM
55.5
TRIFLUOROACETIC ACID
163.8   115.7
200   150   100   50   0 PPM

^{1}H-NMR

MONOSUBSTITUTED ALKANES

^{1}H-Chemical Shifts in Monosubstituted Alkanes

(δ in ppm relative to TMS)

	Substituent	Methyl	Ethyl		n-Propyl			Isopropyl		t-Butyl	Additional Data
		$-CH_3$	$-CH_2$	$-CH_3$	$-CH_2$	$-CH_2$	$-CH_3$	$-CH$	$-CH_3$	$-CH_3$	
	$-H$	0.23	0.86	0.86	0.91	1.33	0.91	1.33	0.91	0.89	H20
	$-CH=CH_2$	1.71	2.00	1.00						1.02	H205
C	$-C\equiv CH$	1.80	2.16	1.15	2.10	1.50	0.97	2.59	1.15	1.22	H225
	$-phenyl$	2.35	2.63	1.21	2.59	1.65	0.95	2.89	1.25	1.32	H30
H	$-F$ [1]	4.27	4.36	1.24						1.34	
A	$-Cl$	3.06	3.47	1.33	3.47	1.81	1.06	4.14	1.55	1.60	H45
L	$-Br$	2.69	3.37	1.66	3.35	1.89	1.06	4.21	1.73	1.76	
	$-I$	2.16	3.16	1.88	3.16	1.88	1.03	4.24	1.89	1.95	
	$-OH$	3.39	3.59	1.18	3.49	1.53	0.93	3.94	1.16	1.22	H50
	$-O-alkyl$	3.24	3.37	1.15	3.27	1.55	0.93	3.55	1.08	1.24	
	$-OC=C$	3.5	3.7	1.3							H60
O	$-O-phenyl$	3.73	3.98	1.38	3.86	1.70	1.05	4.51	1.31		
	$-OCOCH_3$	3.67	4.05	1.21	3.98	1.56	0.97	4.94	1.22	1.45	
	$-OCO-phenyl$	3.88	4.37	1.38	4.25	1.76	1.07	5.22	1.37	1.58	H140
	$-OSO_2-p-toluyl$	3.70	4.07	1.30	3.94	1.60	0.95	4.70	1.25		
	$-NH_2$	2.47	2.74	1.10	2.61	1.43	0.93	3.07	1.03	1.15	H75
N	$-NHCOCH_3$	2.71	3.21	1.12	3.18	1.55	0.96	4.01	1.13	1.28	H150
	$-NO_2$	4.29	4.37	1.58	4.28	2.01	1.03	4.44	1.53	1.59	H90

[1] For $^{19}C-^1H$-coupling see p. H355.

Substituent	Methyl	Ethyl		n-Propyl			Isopropyl		t-Butyl	Additional Data
	$-CH_3$	$-CH_2$	$-CH_3$	$-CH_2$	$-CH_2$	$-CH_3$	$-CH$	$-CH_3$	$-CH_3$	
-SH	2.00	2.44	1.31	2.46	1.57	1.02	3.16	1.34	1.43	H95
-S-alkyl	2.09	2.49	1.25	2.43	1.59	0.98	2.93	1.25	1.39	
-S-S-alkyl	2.30	2.67	1.35	2.63	1.71	1.03			1.32	H110
-CHO	2.20	2.46	1.13	2.42	1.67	0.97	2.39	1.13	1.07	H120
$-COCH_3$	2.09	2.47	1.05	2.32	1.56	0.93	2.54	1.08	1.12	
-CO-phenyl	2.55	2.92	1.18	2.86	1.72	1.02	3.58	1.22		H125
-COOH	2.08	2.36	1.16	2.31	1.68	1.00	2.56	1.21	1.23	H135
$-COOCH_3$	2.01	2.28	1.12	2.22	1.65	0.98	2.48	1.15	1.16	H140
$-CONH_2$	2.02	2.23	1.13	2.19	1.68	0.99	2.44	1.18	1.22	H150
-C=N-OH	1.9									H175
-CN	1.98	2.35	1.31	2.29	1.71	1.11	2.67	1.35	1.37	H180

¹H-NMR

SUBSTITUTED ALKANES

Estimation of the Chemical Shift in Polysubstituted Alkanes

(δ in ppm relative to TMS)

$$\delta_{CH_2R_1R_2} = 1.25 + \sum_{1}^{2} a_i \qquad \delta_{CHR_1R_2R_3} = 1.50 + \sum_{1}^{3} a_i$$

	Substituent	a_i
C	-alkyl	0.0
	-C=C-	0.8
	-C≡C-	0.9
	-phenyl	1.3
H	-Cl	2.0
A	-Br	1.9
L	-I	1.4
	-OH	1.7
	-O-alkyl	1.5
O	-O-phenyl	2.3
	-OCO-alkyl	2.7
	-OCO-phenyl	2.9
	-NH$_2$	1.0
N	-N-alkyl$_2$	1.0
	-NO$_2$	3.0
S	-S-alkyl	1.0
	-CHO	1.2
O	-CO-alkyl	1.2
‖	-COOH	0.8
C	-COO-alkyl	0.7
	-CN	1.2

Example:

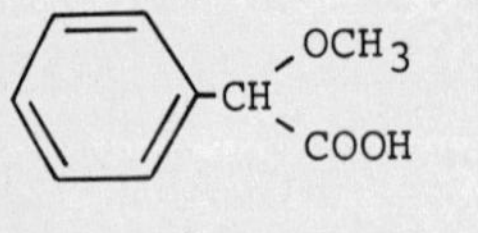

base value:	1.5
-O-alkyl	1.5
-COOH	0.8
-phenyl	1.3
estimated:	5.1
determined:	4.8

H15

Coupling in Aliphatic Compounds

for coupling in alicyclics see p. H185

Geminal coupling

$$J_{HCH} = -8 \cdots\cdots -18 \text{ Hz}$$

Electronegative substituents cause a decrease in $|J_{gem}|$ while an sp^2- or sp-hybridized carbon atom as the substituent causes an increase:

Compound	J_{gem} [Hz]	Compound	J_{gem} [Hz]
CH_4	-12.4	CH_3COCH_3	-14.9
CH_3Cl	-10.8	CH_3-phenyl	$\sim$ -14.3
CH_2Cl_2	-7.5	CH_3CN	-16.9
CH_3OH	-10.8	$CH_2(CN)_2$	-20.3

Vicinal coupling

- for free rotation $\qquad J_{HCCH} = \sim 7$ Hz

- for fixed conformation $\qquad J_{HCCH} = 0-18$ Hz

The nature of the substituent has little influence on the vicinal coupling constants if free rotation is possible:

$$CH_3CH_2-Li \qquad J_{vic} = 8.4 \text{ Hz}$$

$$CH_3CH_2-F \qquad J_{vic} = 6.9 \text{ Hz}$$

The vicinal coupling constants are mainly dependent on the dihedral angle ϕ:

$$J = J^0 \cos^2 \phi - 0.3 \qquad 0° < \phi < 90°$$

$$J = J^{180} \cos^2 \phi - 0.3 \qquad 90° < \phi < 180°$$

^{1}H-NMR

The same relationship between dihedral angle and vicinal coupling
constant is frequently observed in substituted alkanes. Both J^0
and J^{180} can be affected by the substituents. The following figure
gives the values calculated for various values of J^0 and J^{180} as
a function of the dihedral angle.

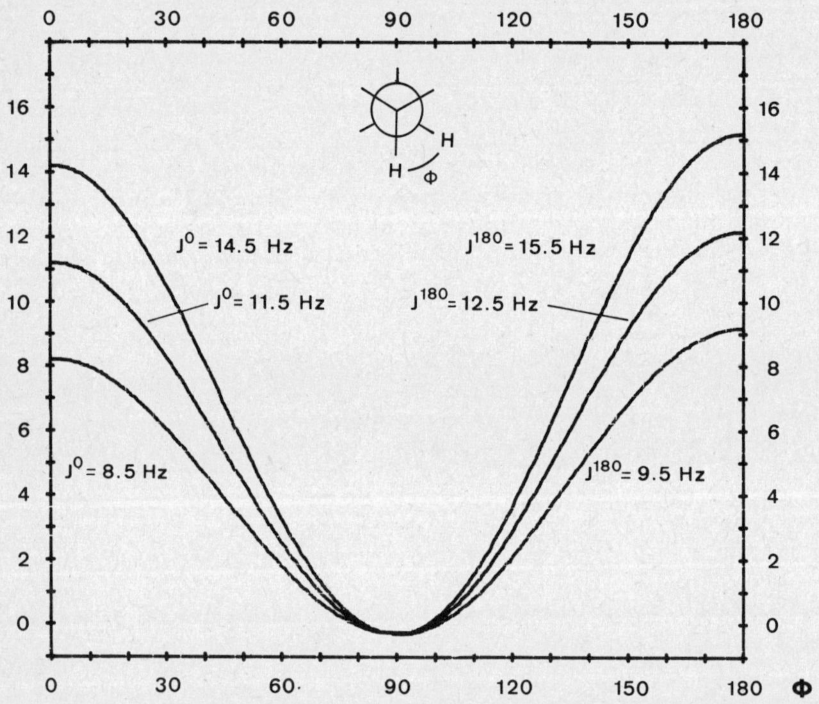

An excellent discussion of the scope and limitations of conformational analysis based on such relationships can be found on p. 281
of Jackman and Sternhell (see p. A15).

Coupling constants across more than three bonds ("long range"
coupling) in saturated hydrocarbons are generally in the order of
<~1 Hz. However, they are of significance for fixed conformations
of condensed, alicyclic systems (see p. H190) and in unsaturated
compounds (see p. H205, H210).

^{1}H-Chemical Shifts in Aromatic Substituted Alkanes

(δ in ppm relative to TMS)

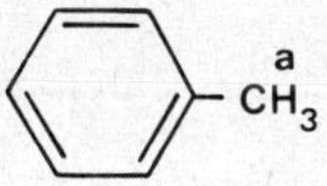

(a) 2.35

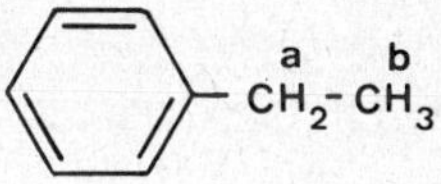

(a) 2.63
(b) 1.21

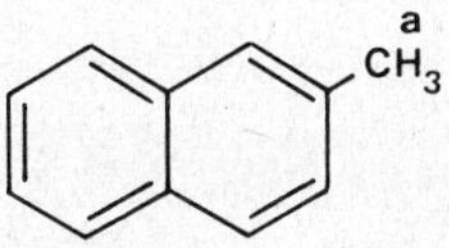

(a) 2.46

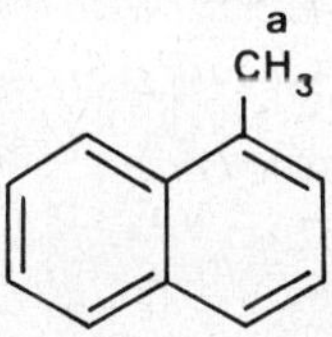

(a) 2.65

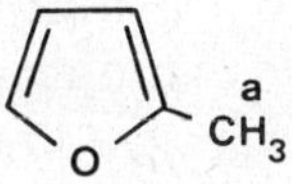

(a) 2.17

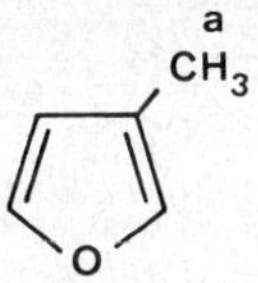

(a) 1.94

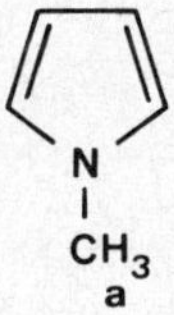

(a) 3.50

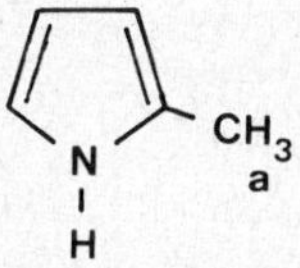

(a) 2.16

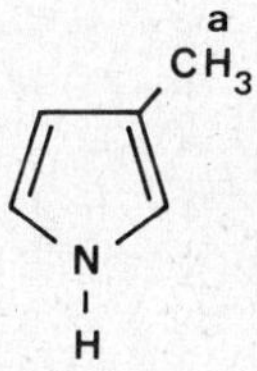

(a) 2.05

¹H-NMR

AROMATIC SUBSTITUTED ALKANES

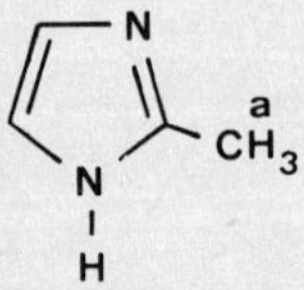

(a) 2.42

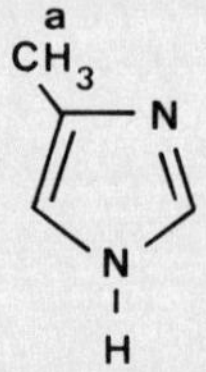

(a) 2.27

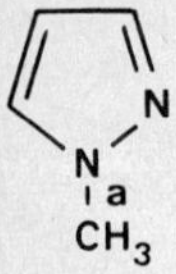

(a) 3.80

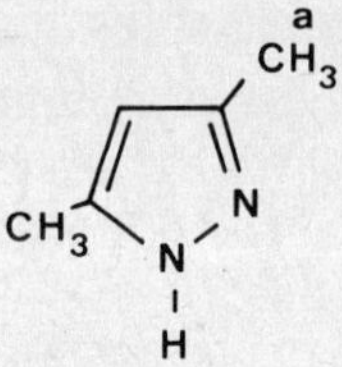

(a) 2.79

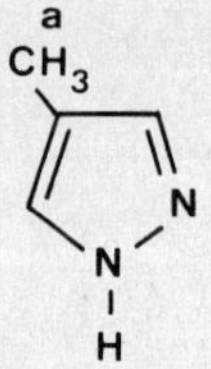

(a) 2.05

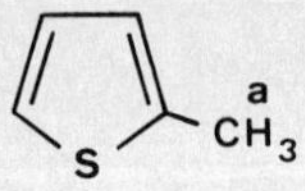

(a) 2.41

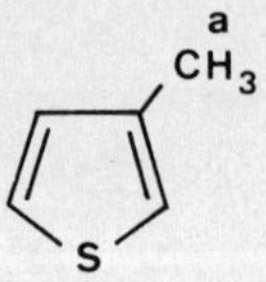

(a) 2.21

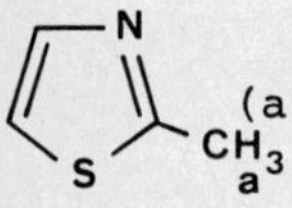

(a) 2.74

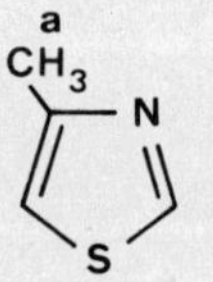

(a) 2.47

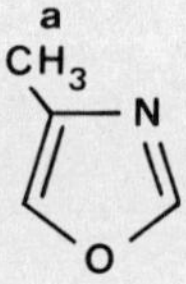

(a) 2.18

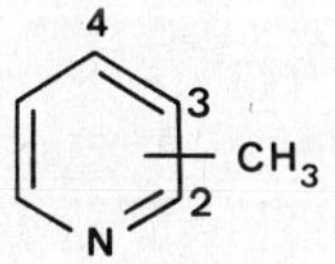

position	δ_{CH_3}
2	2.55
3	2.32
4	2.37

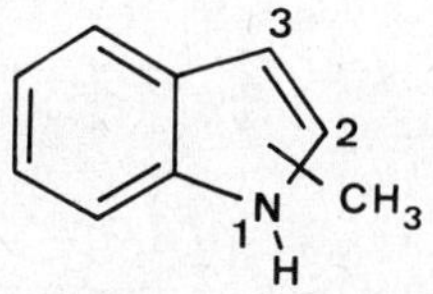

position	δ_{CH_3}
1	3.60
2	3.20
3	2.30

^{1}H-NMR

HALOALKANES

<u>^{1}H-Chemical Shifts in Halogenated Methane and Ethane Derivatives</u>

(δ in ppm relative to TMS)

Compound type	Chemical shift for X =			
	$F^{1)}$	Cl	Br	I
CH_3X	4.27	3.06	2.69	2.16
CH_2X_2	5.45	5.33	4.94	3.90
CHX_3	6.49	7.24	6.82	4.91
CH_2-X	4.36	3.47	3.37	3.16
CH_3	1.24	1.33	1.66	1.88
CHX_2		5.8	5.86	
CH_3		2.1	2.47	
XCH_2CH_2X		3.7	3.63	3.7

$^{1)}$ For ^{19}F-^{1}H-coupling see p. H355.

Chemical Shift for Hydroxyl Protons

(δ in ppm relative to TMS)

Aliphatic (and alicyclic) alcohols: 0.5 - 5.0
Phenols: 5.0 - 8.0

Hydrogen bonds strongly deshield hydroxyl protons (in the enol form
of β-dicarbonyl compounds to over 15). The position of the signal
may be strongly dependent on the experimental conditions. If the
substance contains, in addition, acidic protons (-OH, -COOH, H_2O)
only one signal in the average position is generally observed
because of rapid exchange.

In dimethyl sulfoxide as solvent this exchange is often so slow that
isolated signals are observed (most important exceptions are sub-
stances containing strong acids or amines). In this solvent the
chemical shifts of hydroxyl protons are characteristic (see p. H55).

^{1}H-NMR

ALIPHATIC ALCOHOLS

Chemical Shifts of Hydroxyl Protons in Dimethyl Sulfoxide as the Solvent (δ in ppm relative to TMS)

CH_3CH_2OH	4.5	cyclohexanol (equatorial OH)	4.0 $\cdots$ 4.5
CCl_3CH_2OH	6.8	cyclohexanol (axial OH)	3.8 $\cdots$ 4.2
C_6H_5–CH_2OH	5.2	C_6H_5–OH	9.3
$(C_6H_5)_2CHOH$	5.6	NO_2–C_6H_4–OH	11.0
$(C_6H_5)_3COH$	6.4	3-hydroxypyridine	9.8

Chemical Shifts for Protons in the Vicinity of Hydroxyl Groups (δ in ppm relative to TMS)

For alicyclic alcohols see p. H195.
For phenols see p. H255.

$\overset{a}{CH_3}OH$ (a) 3.39

$\overset{b}{CH_3}\overset{a}{CH_2}OH$ (a) 3.59   (b) 1.18

$\overset{c}{CH_3}\overset{b}{CH_2}\overset{a}{CH_2}OH$ (a) 3.49   (b) 1.53   (c) 0.93

$(\overset{b}{CH_3})_2\overset{a}{CHOH}$ (a) 3.94   (b) 1.16

$(\overset{a}{CH_3})_3COH$ (a) 1.22

Coupling of Hydroxyl Protons (J in Hz)

The coupling between hydroxyl protons and the protons on the
adjacent carbon atoms are normally not observable in the spectrum
because of the fast exchange of the hydroxyl protons. However,
in very pure (acid-free) solutions or in dimethyl sulfoxide as
the solvent the exchange is sufficiently slow that the H-O-C-H
couplings become visible. They exhibit a dependence on the
conformation analogous to that shown by the H-C-C-H coupling
(see p. H20). In case of free rotation:

$$J_{HOCH} = \sim 5$$

In cyclohexanols the coupling constants for axial and equatorial
hydroxyls measured in dimethyl sulfoxide differ significantly:

$$J_{HOCH} = 4.2 \cdots 5.7$$

$$J_{HOCH} = 3.0 \cdots 4.2$$

^{1}H-Chemical Shifts in Aliphatic Ethers (δ in ppm relative to TMS)

$\overset{a}{CH_3}$O-alkyl	(a) 3.24	$\overset{b}{CH_3}\overset{a}{CH_2}$O-alkyl	(a) 3.37 (b) 1.15
$\overset{a}{CH_3}$O-C=C	(a) 3.5	$\overset{b}{CH_3}\overset{a}{CH_2}$O-C=C	(a) 3.7 (b) 1.3
$\overset{a}{CH_3}$O- (H b)	(a) 3.73 $|J|_{ab} = 0 - 0.8$	$\overset{b}{CH_3}\overset{a}{CH_2}$O-	(a) 3.98 (b) 1.38
$CH_3\overset{a}{O}CH_3$	(a) 3.3	$\overset{a}{CH_2}\overset{b}{(OCH_3)_2}$	(a) 3.3 (b) 4.5
$\overset{b}{CH}\overset{a}{(OCH_3)_3}$	(a) 3.3 (b) 5.1	$(CH_3)_2\overset{b}{CH}\overset{a}{O}$-alkyl	(a) 3.55 (b) 1.08

^{1}H-Chemical Shift (δ in ppm relative to TMs) and Coupling Constants (J in Hz) in Non-Aromatic Cyclic Ethers

For heteroaromatic systems see p. H265 - H350.

(a) 2.54

geminal coupling: J_{gem} = 6

vicinal couplings: J_{cis} = 4.5
J_{trans} = 3.1

(a,c) 4.73
(b) 2.72

geminal couplings: (a) J_{gem} = -5.8
(b) J_{gem} = -11.0

vicinal couplings: J_{cis} = 8.7
J_{trans} = 6.6

"long-range" coupling: $|J|_{ac}$ < 0.3

(a) 3.75
(b) 1.85

(a) 4.20
(b) 2.53
(c) 4.82
(d) 6.22

J_{ab} = 9.3 $\qquad$ J_{bd} = 2.6

J_{bc} = 2.5 $\qquad$ J_{cd} = 2.6

(a,d) 4.43
(b,c) 5.78

J_{ab} = $\sim$ 1

J_{ac} = $\sim$-1

J_{bc} = $\sim$ 6

(a) 3.9   geminal coupling   (a) $J_{gem} = \sim-7.5$
(b) 4.9   vicinal couplings $J_{cis} = 7.3$
                                    $J_{trans} = 6.0$

(a) 5.90   (a) $|J|_{gem} = \sim1.5$

(a) 3.6
(b,c) $\sim$1.6

(a) 4.0   (c) $\sim$1.9   (e) 6.4
(b) $\sim$1.9   (d) 4.6

(a,e) 6.17 $J_{ab} = 7.0$ $J_{ae} = 1.5$
(b,d) 4.63 $J_{ac} = 1.7$ $J_{bc} = 3.4$
(c)   2.66

(a,d) 6.35 $J_{ab} = 5.9$
(b,c) 7.71 $J_{ac} = 0.4$
             $J_{ad} = 2.7$
             $J_{bc} = 1.1$

(a) 3.71

(a) 4.70
(b) 3.80
(c) 1.68

(a) 5.00

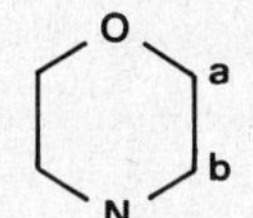

(a) 3.67
(b) 2.87
(c) 1.92

(a) 3.88
(b) 2.57

^{1}H-NMR

ALIPHATIC AMINES

^{1}H-Chemical Shifts in Aliphatic Amines (δ in ppm relative to TMS)

alkyl–NH$_2$ (a) (alkyl)$_2$–NH (a)	(a) 0.5–4.0
alkyl–NH$_3^+$ (a) (alkyl)$_2$–NH$_2^+$ (a) (alkyl)$_3$–NH$^+$ (a)	(a) ∿6–9

phenyl–NH$_2$ (a) phenyl–NH–alkyl (a) phenyl–NH–phenyl (a)	(a) 2.5–5.0
phenyl–NH$_3^+$ (a) phenyl–NH$_2^+$–alkyl (a) phenyl–NH$^+$–(alkyl)$_2$ (a)	(a) ∿6–9

CH$_3$NH$_2$ (a)   (a) 2.47

(CH$_3$)$_2$NH (a)   (a) ∿2.3

CH$_3$CH$_2$NH$_2$ (b, a)   (a) 2.74 (b) 1.10

CH$_3$–NH–cyclohexyl (a)   (a) 2.44

(CH$_3$)$_2$CHNH$_2$ (b, a)   (a) 3.07 (b) 1.03

CH$_3$–NH–phenyl (a, b)   (a) 2.78 (b) 3.55

CH$_3$NH–C$_6$H$_4$–NO$_2$ (a, b)   (a) 2.93 (b) 4.62

(CH$_3$)$_2$–N–cyclohexyl (a)   (a) 2.25

(CH$_3$)$_4$N$^+$ (a)   (a) 3.3

CH$_3$–N–piperidyl (a)   (a) 2.21

(CH$_3$CH$_2$)$_4$N$^+$ (b, a)   (a) 3.25 (b) 1.25

(CH$_3$)$_2$–N–phenyl (a)   (a) 2.85

(CH$_3$)$_3$N$^+$–phenyl (a)   (a) 3.72

(CH$_3$)$_2$–N–C$_6$H$_4$–NO$_2$ (a)   (a) 3.09

Coupling in Amines (J in Hz)

Coupling with Amine Protons

The H-N-C-H coupling is not observable because of the rapid exchange of the NH-protons. However, for a CH-NH-C= moiety (enamines, aromatic amines, amides, etc.) splitting is often observed. The coupling constant exhibits a dependence on the conformation analogous to that shown by vicinal H-C-C-H couplings (see p. H2O). In case of free rotation:

$$J_{H-C-N-H} = 5 - 6$$

In trifluoroacetic acid as the solvent the exchange of the ammonium protons is slowed to such an extent that the coupling of $H-N^+-C-H$ generally becomes observable; in case of free rotation:
$$J_{H-N^+-C-H} = 5 - 6.$$

Coupling with ^{14}N

Generally $^{14}N-^1H$-couplings are not observable because of the quadrupole relaxation of the nitrogen nucleus, even if the exchange of the amino protons is slow (as in weakly basic amines, amides and ammonium ions). For the same reason $^{14}N-CH-CH$ couplings are generally not observable. In ammonium compounds of high symmetry the contribution of the quadrupole relaxation is, however, so small that the $^{14}N-^1H$ and the $^{14}N-CH-CH$ couplings become observable:

$$
\begin{array}{lll}
^+NH_4 & |J|_{^{14}N-H} & = 52.8 \\
^+N(CH_2CH_2CH_3)_4 & |J|_{^{14}N-CH} & = 0.5 \\
 & |J|_{^{14}N-C-CH} & = 1.6 \\
 & |J|_{^{14}N-C-C-CH} & = 0
\end{array}
$$

In many cases the combination of quadrupole relaxation and exchange of NH-protons is not sufficiently large to completely eliminate the large $^{14}N-^1H$ coupling across one bond. In such a case the signal for the NH-protons is often broad (for example in ammonium compounds, amides, NH-protons of heteroaromatics, etc.). However, such a line broadening has no effect on the H-N-C-H coupling.

^{1}H-NMR

CYCLIC AMINES

<u>^{1}H-Chemical Shifts (δ in ppm relative to TMS) and Coupling Constants</u>
<u>(J in Hz) in Non-Aromatic Cyclic Amines</u>

For heteroaromatic systems see p. H265 - H350.

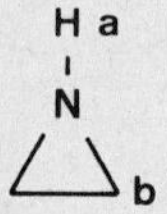

(a) 0.9
(b) 1.61

geminal coupling: $J_{gem} = {\sim}1$
vicinal couplings: $J_{cis} = {\sim}6$
                       $J_{trans} = {\sim}4$

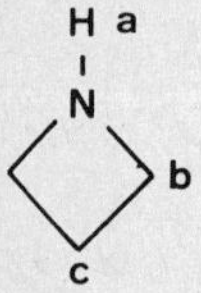

(a) 2.38
(b) 3.54
(c) 2.23

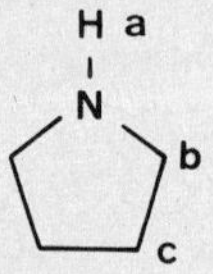

(a) 2.01
(b) 2.75
(c) 1.59

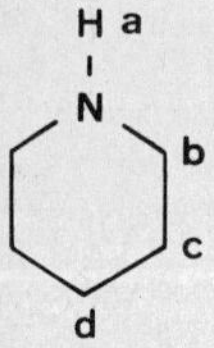

(a) 1.84
(b) 2.74
(c) 1.5
(d) 1.5

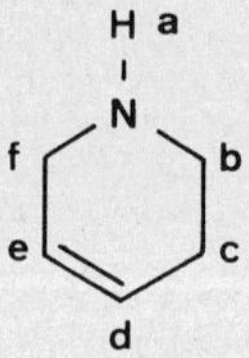

(a) 1.63
(b) 2.95
(c) 2.07
(d)
(e) 5.72 & 5.77
(f) 3.33

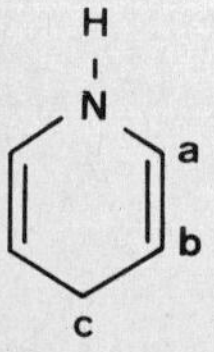

(a) 5.73
(b) 4.42
(c) 3.15

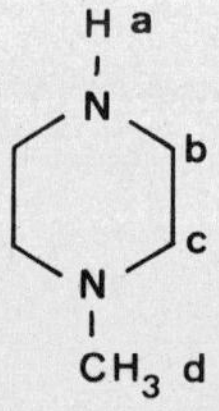

(a) 2.12
(b) 2.88
(c) 2.37
(d) 2.27

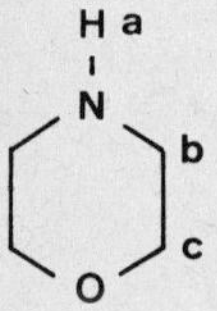

(a) 1.92
(b) 2.87
(c) 3.67

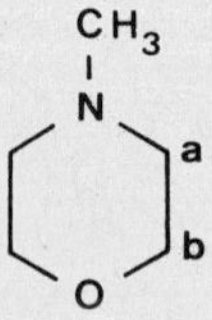

(a) 2.32
(b) 3.62

^{1}H-Chemical Shifts in Aliphatic Nitro Compounds

(δ in ppm relative to TMS)

$\overset{a}{C}H_3NO_2$ (a) 4.29

$\overset{b}{C}H_3\overset{a}{C}H_2NO_2$ (a) 4.37   (b) 1.58

$\overset{c}{C}H_3\overset{b}{C}H_2\overset{a}{C}H_2NO_2$ (a) 4.28   (b) 2.01   (c) 1.03

$(\overset{b}{C}H_3)_2\overset{a}{C}HNO_2$ (a) 4.44   (b) 1.53

$(\overset{a}{C}H_3)_3CNO_2$ (a) 1.59

$NO_2-CH_2-CH_2-CH_2-CH_2-CH_2-CH_3$:
CH_2 4.3, CH_2 2.0, CH_2/CH_2 ~1.4, CH_3 0.9

^{1}H-Chemical Shifts in Aliphatic N-Nitroso, Azo and Azoxy Compounds

(δ in ppm relative to TMS)

$\overset{a}{C}H_3$ / $\overset{b}{C}H_3$ $N^+=N$ O^- (a) 2.96   (b) 3.76

$(\overset{b}{C}H_3)_2\overset{a}{C}H$ / $(\overset{d}{C}H_3)_2\overset{c}{C}H$ $N^+=N$ O^- (a) 4.89   (b) 1.15   (c) 4.26   (d) 1.52

$\delta_{cis} < \delta_{trans}$ for αCH_3-, αCH_2- and βCH_3-protons

$\delta_{cis} > \delta_{trans}$ for αCH-protons

$\overset{a}{C}H_3-N=N-CH_3$ (a) 3.7

$\overset{a}{C}H_3-N=N(\rightarrow O)-\overset{b}{C}H_3$ (a) 4.16   (b) 3.16

$\overset{a}{C}H_3-N=N-C_6H_5$ (a) 3.4

$(\overset{a}{C}H_3)_3C-N=N(\rightarrow O)-\overset{b}{C}(CH_3)_3$ (a) 1.48   (b) 1.28

ALIPHATIC THIOLS,
THIOETHERS

<u>¹H-Chemical Shifts (δ in ppm relative to TMS) and Coupling Constants</u>
<u>(J in Hz) in Aliphatic Thiols</u>

$$\text{alkyl-}\overset{a}{\text{SH}} \qquad \text{(a) 1-2}$$

$$\bigcirc\!\!-\overset{a}{\text{SH}} \qquad \text{(a) 2-4}$$

The exchange with other -SH, -COOH or -OH protons is generally so
slow that the chemical shift is characteristic and the coupling with
the -SH protons becomes observable (J = 5 - 9 Hz for free rotation).

$\overset{a}{CH_3}SH$	(a) 2.00		$(\overset{c}{CH_3})_2\overset{b}{CH}\overset{a}{SH}$	(a) 1.56
				(b) 3.16
$\overset{b}{CH_3}\overset{a}{CH_2}SH$	(a) 2.44			(c) 1.34
	(b) 1.31		$(\overset{a}{CH_3})_3CSH$	(a) 1.43
$\overset{c}{CH_3}\overset{b}{CH_2}\overset{a}{CH_2}SH$	(a) 2.46		$HS\overset{c}{CH_2}\overset{b}{CH_2}\overset{a}{CH_2}SH$	(a) 1.35
	(b) 1.57			(b) 2.68
	(c) 1.02			(c) 1.88

<u>¹H-Chemical Shifts in Aliphatic Thioethers</u>
(δ in ppm relative to TMS)

$\overset{a}{CH_3}S\text{-alkyl}$	(a) 2.09		$\overset{c}{CH_3}\overset{b}{CH_2}\overset{a}{CH_2}S\text{-alkyl}$	(a) 2.43
$\overset{a}{CH_3}S\text{-}C=C$	(a) 2.25			(b) 1.59
				(c) 0.98
$\overset{a}{CH_3}S\!\!-\!\!\bigcirc$	(a) 2.47		$(\overset{b}{CH_3})_2\overset{a}{CH}S\text{-alkyl}$	(a) 2.93
$\overset{b}{CH_3}\overset{a}{CH_2}S\text{-alkyl}$	(a) 2.49			(b) 1.25
	(b) 1.25		$(\overset{a}{CH_3})_3CS\text{-alkyl}$	(a) 1.39

^{1}H-Chemical Shifts (δ in ppm relative to TMS) and Coupling Constants (J in Hz) in Non-Aromatic Cyclic Thioethers and Sulfones

For heteroaromatic systems see p. H265 - H350.

(a) 2.27

geminal coupling: $J_{gem} = 0$
vicinal couplings: $J_{cis} = 7.2$
$J_{trans} = 5.7$

(a) 3.21
(b) 2.94

geminal couplings: (a) $J_{gem} = -8.7$
(b) $J_{gem} = -11.7$
vicinal couplings: $J_{cis} = 8.9$
$J_{trans} = 6.3$
"long-range" couplings: $J_{cis} = 1.2$
$J_{trans} = -0.2$

(a) 2.75
(b) 1.88

(a) 3.00
(b) 2.23

(a) 3.67
(b) 5.81

(a) 3.74
(b) 6.08

(a) 3.08
(b) 2.62
(c) 5.48
(d) 6.06

$J_{ab} = 9.2$
$J_{bc} = 2.5$
$J_{cd} = 6.1$
$J_{bd} = 2.2$

^{1}H-NMR

CYCLIC THIOETHERS

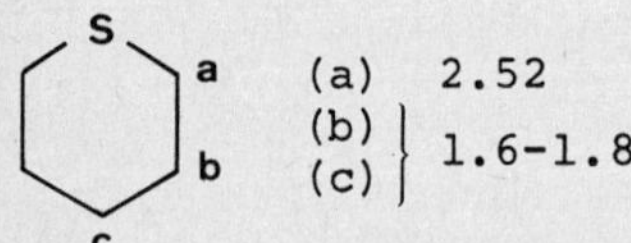

(a)	2.52
(b)	} 1.6–1.8
(c)	

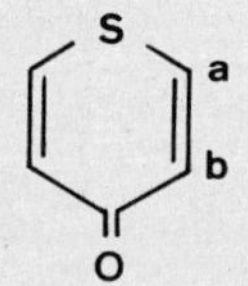

(a)	7.89
(b)	7.09

(a,e)	5.97
(b,d)	5.55
(c)	2.84

$J_{ab} = 10.0$ $J_{bd} = 0.0$
$J_{bc} = 3.9$ $J_{ae} = 2.9$
$J_{ac} = 1.1$ $J_{ad} = 0.0$

(a)	3.19
(b)	
(c)	} 5.5–6.2
(d)	
(e)	

(a)	4.00
(b)	
(c)	} 6.3–6.7
(d)	
(e)	

(a)	2.7
(b)	1.9

(a)	4.2

(a)	2.57
(b)	3.88

^{1}H-Chemical Shifts in Aliphatic S-Compounds

(δ in ppm relative to TMS)

For thiols and thioethers see p. H95 - H105.

Disulfides

$\overset{a}{CH_3}SS\text{-alkyl}$ (a) 2.30

$\overset{b}{CH_3}\overset{a}{CH_2}SS\text{-alkyl}$ (a) 2.67
           (b) 1.35

$\overset{c}{CH_3}\overset{b}{CH_2}\overset{a}{CH_2}SS\text{-alkyl}$ (a) 2.63
           (b) 1.71
           (c) 1.03

$(\overset{a}{CH_3})_3CSS\text{-alkyl}$ (a) 1.32

Sulfoxides

$\overset{a}{CH_3}SOCH_3$ (a) 2.50

Sulfones

$\overset{a}{CH_3}SO_2CH_3$ (a) 2.84

$\overset{a}{CH_3}SO_2\overset{b}{CH_2}\overset{c}{CH_3}$ (a) 2.80
           (b) 2.94
           (c) 1.47

$\overset{a}{CH_3}SO_2\overset{b}{CH}=\overset{c}{CH_2}$ (a) 2.62
           (b) 6.70
           (c) cis   6.13
               trans 5.95

$(\overset{a}{CH_3})_3CSO_2C(CH_3)_3$ (a) 1.44

Derivatives of Sulfinic and Sulfonic Acids

$\overset{a}{CH_3}SON(\overset{b}{CH_3})_2$ (a) 2.50
           (b) 2.68

$\overset{a}{CH_3}SO_2\text{-O-alkyl}$ (a) 3.0

$\overset{a}{CH_3}SO_2NH\text{—}\langle\text{phenyl}\rangle$ (a) 2.82

$\overset{a}{CH_3}SO_2Cl$ (a) 3.6

$\text{alkyl-}SO_2\overset{a}{OH}$

$\langle\text{phenyl}\rangle\text{-}SO_2\overset{a}{OH}$
     (a) 11-12

(a) 3.19

(b) 2.44

(c) 4.27

For esters of sulfonic acids see also p. H140, H145.

^{1}H-NMR

Esters of Sulfuric Acid and Sulfurous Acid

$$\overset{a}{CH_3}OSO-OCH_3 \qquad (a)\ 3.58$$

$$\overset{a}{CH_3}OSO_2-OCH_3 \qquad (a)\ 3.94$$

(a) 4.68

Thiocarboxylic Acids and Derivatives

$$\overset{a}{CH_3}\overset{b}{COSH} \qquad \begin{array}{l}(a)\ 2.4 \\ (b)\ 4.7\end{array}$$

$$\overset{a}{CH_3}\overset{b}{COSCH_3} \qquad \begin{array}{l}(a)\ 2.30 \\ (b)\ 2.27\end{array}$$

(a) 7.1
(b) 6.9
(c) 6.4
(d) 7.7

(a) 6.47 $\qquad J_{ab} = 5.9$
(b) 7.84 $\qquad J_{ac} = 2.0$
(c) 4.29 $\qquad J_{bc} = 2.8$

Thiocyanates, Isothiocyanates

$$\overset{a}{CH_3}SCN \qquad (a)\ 2.61$$

$$\overset{a}{CH_3}NCS \qquad (a)\ 3.37$$

$$\overset{b}{CH_3}\overset{a}{CH_2}SCN \qquad \begin{array}{l}(a)\ 2.98 \\ (b)\ 1.52\end{array}$$

$$\overset{b}{CH_3}\overset{a}{CH_2}NCS \qquad \begin{array}{l}(a)\ 3.64 \\ (b)\ 1.40\end{array}$$

$$(CH_3)_2\overset{a}{CH}SCN \qquad (a)\ 3.48$$

$$(CH_3)_2\overset{a}{CH}NCS \qquad (a)\ 3.98$$

^{1}H-Chemical Shifts (δ in ppm relative to TMS) and Coupling Constants (J in Hz) in Aliphatic Aldehydes

alkyl-$\overset{a}{C}HO$

alkenyl-$\overset{a}{C}HO$

(a) 9.0-10.1

$\bigcirc\!\!-\overset{a}{C}HO$ unsubstituted in the ortho-position: (a) 9.6-10.2
substituted in the ortho-position: (a) 10.2-10.5

Couplings:

$\underset{\overset{|}{H}}{\overset{}{\geq}}C\!-\!\underset{\overset{|}{H}}{C}\!-\!C\!=\!O$ $|J|_{HC-CHO} = 0...3$ (depending on conformation)

$|J|_{ab} = \sim 8$

$|J|_{ac} = \sim 0.3$

$|J|_{ad} = \sim 0.1$

$|J|_{ab} = \sim 0.2$

$|J|_{ac} = 0$

$|J|_{ac'} = 0.4 - 1.0$

$|J|_{ad} = 0.1$

H-CHO $\delta = 9.60$
$|J| = 42.4$

$\overset{b}{C}H_3\overset{a}{C}HO$ (a) 9.80
(b) 2.20
$J_{ab} = 3$

$\overset{b}{C}H_3\overset{a}{C}H_2CHO$ (a) 2.46
(b) 1.13

$\overset{d}{C}H_3\overset{c}{C}H_2\overset{b}{C}H_2\overset{a}{C}HO$ (a) 9.74
(b) 2.42
(c) 1.67
(d) 0.97
$J_{ab} = 2$

$(\overset{b}{C}H_3)_2\overset{a}{C}HCHO$ (a) 2.39
(b) 1.13

$(\overset{a}{C}H_3)_3CCHO$ (a) 1.07

^{1}H-NMR

ALIPHATIC KETONES

^{1}H-Chemical Shifts (δ in ppm relative to TMS) and Coupling Constants
(J in Hz) in Aliphatic Ketones

$\overset{a}{C}H_3COCH_3$ (a) 2.09 $\overset{a}{C}H_3CO$—⟨phenyl⟩   (a) 2.55

$\overset{a}{C}H_3COC=C$ (a) 2.3

$\overset{b}{C}H_3\overset{a}{C}H_2COCH_3$ (a) 2.47   (b) 1.05 $\overset{b}{C}H_3\overset{a}{C}H_2CO$—⟨phenyl⟩   (a) 2.92   (b) 1.18

$\overset{c}{C}H_3\overset{b}{C}H_2\overset{a}{C}H_2COCH_3$ (a) 2.32   (b) 1.56   (c) 0.93 $\overset{c}{C}H_3\overset{b}{C}H_2\overset{a}{C}H_2CO$—⟨phenyl⟩   (a) 2.86   (b) 1.72   (c) 1.02

$(\overset{b}{C}H_3)_2\overset{a}{C}HCOCH_3$ (a) 2.54   (b) 1.08 $(\overset{b}{C}H_3)_2\overset{a}{C}HCO$—⟨phenyl⟩   (a) 3.58   (b) 1.22

$(\overset{a}{C}H_3)_3CCOCH_3$ (a) 1.12

$\overset{a}{C}H_3CO\overset{b}{C}H_2COCH_3$ (a) 2.17   (b) 3.62

$J_{HCCOCH} = 0\text{--}0.5$

larger if conformation fixed ("W-effect"):

$J_{ab} = 1$

H125

<u>¹H-Chemical Shifts (δ in ppm relative to TMS) and Coupling Constants (J in Hz) in Cyclic Ketones</u>

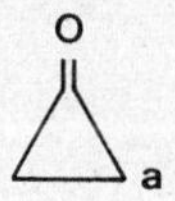

(a) 1.65

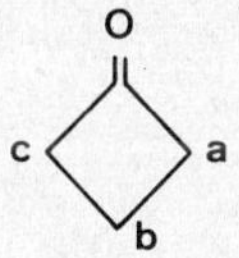

(a,c) 3.03
(b)   1.96
$J_{ac(cis)}$ = 1.9
$J_{ac(trans)}$ = 1.0

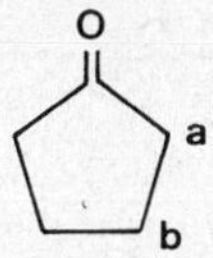

(a) 2.06
(b) 2.02

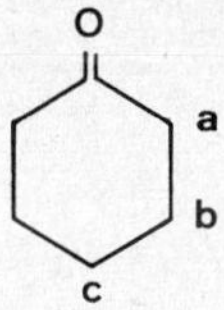

(a)   2.22
(b,c) ca. 1.8

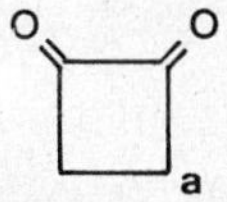

(a) 3.0

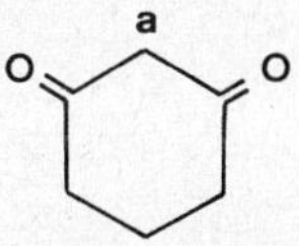

(a) 3.4

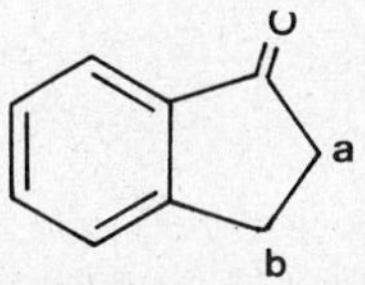

(a) 2.60
(b) 3.08

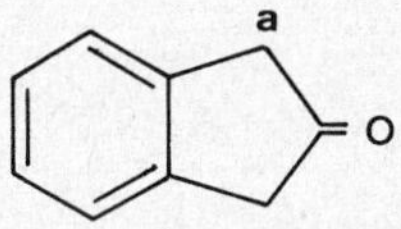

(a) 3.50

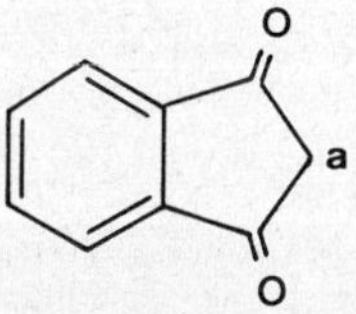

(a) 3.23

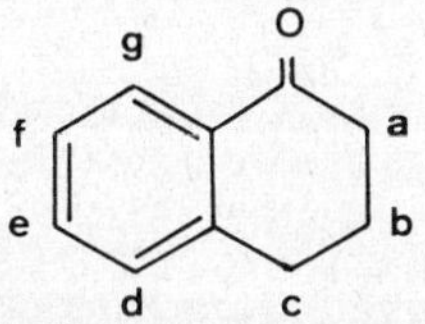

(a)   2.58
(b)   2.18
(c)   2.92
(d,e,f) 7.0-7.4
(g)   8.0

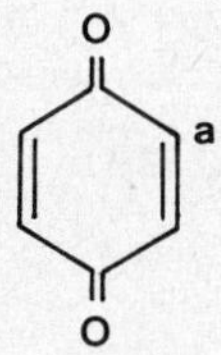

(a) 6.78

4-oxopyran: see p. H70.

4-oxo-1-thiopyran: see p. H105.

^{1}H-NMR

CARBOXYLIC ACIDS

[1]H-Chemical Shifts in Aliphatic Carboxylic Acids

(δ in ppm relative to TMS)

$$\left.\begin{array}{l} \text{alkyl-}\overset{a}{C}OOH \\[1em] \text{phenyl-}\overset{a}{C}OOH \end{array}\right\} \quad \text{(a) } 10\text{-}13$$

The position of the signals depends on the solvent, the concentration and the presence of other exchangeable protons (e.g. alcohols, water, etc.).

$\overset{a}{C}H_3COOH$	(a) 2.08	$(\overset{b}{C}H_3)_2\overset{a}{C}HCOOH$	(a) 2.56 (b) 1.21
$\overset{b}{C}H_3\overset{a}{C}H_2COOH$	(a) 2.36 (b) 1.16	$(\overset{a}{C}H_3)_3CCOOH$	(a) 1.23
$\overset{c}{C}H_3\overset{b}{C}H_2\overset{a}{C}H_2COOH$	(a) 2.31 (b) 1.68 (c) 1.00		

$HOOC\overset{a}{C}H_2COOH \qquad \text{(a) } 3.4$

$HOOC\overset{a}{C}H_2\overset{a}{C}H_2COOH \qquad \text{(a) } 2.6$

^{1}H-Chemical Shifts (δ in ppm relative to TMS) and Coupling Constants (J in Hz) in Aliphatic Esters

Formates

$\overset{a}{H}COOCH_3$	(a) 8.03	$\overset{a}{H}COOC-C-$ with $\overset{b}{H}\ \overset{c}{H}$ $J_{ab} = \sim -1$, $J_{ac} = \sim 0.5$

$\overset{a}{H}COO$-C=C with $\overset{d}{H}$, $\overset{b}{H}$, $\overset{c}{H}$ (a) 8.1 $J_{ab} = -0.7$, $J_{ac} = 1.6$, $J_{ad} = 0.8$

Acetates

$\overset{a}{C}H_3\overset{b}{C}OOCH_3$ (a) 2.01   (b) 3.67

$\overset{a}{C}H_3COOC=C$ (a) 2.1

$\overset{a}{C}H_3COO-$⟨phenyl⟩   (a) 2.1

Propionates

$\overset{b}{C}H_3\overset{a}{C}H_2COOCH_3$ (a) 2.28   (b) 1.12

Isobutyrates

$(\overset{b}{C}H_3)_2\overset{a}{C}HCOOCH_3$ (a) 2.48   (b) 1.15

Methyl esters

$\overset{a}{C}H_3OCOCH_3$ (a) 3.67 $\overset{a}{C}H_3OCO-$⟨phenyl⟩   (a) 3.88

$\overset{a}{C}H_3OCOC=C$ (a) 3.8 $\overset{a}{C}H_3OSO_2-$⟨phenyl⟩$-CH_3$ (a) 3.70

Methyl ester of:		δ		δ
	boric acid	3.5	perchloric acid	4.3
	silicic acid	3.6	nitric acid	4.2
	carbonic acid	3.8	sulfuric acid	3.94
	phosphoric acid	3.8	sulfurous acid	3.58

^{1}H-NMR

ALIPHATIC ESTERS, LACTONES

Ethyl esters

$$CH_3CH_2OCOCH_3$$ (b,a)
(a) 4.05
(b) 1.21

$$CH_3CH_2OCO\text{-}C_6H_5$$ (b,a)
(a) 4.37
(b) 1.38

$$CH_3CH_2OCOCF_3$$ (b,a)
(a) 4.3

$$CH_3CH_2OSO_2\text{-}C_6H_4\text{-}CH_3$$ (b,a)
(a) 4.07
(b) 1.30

$$CH_3CH_2OCOC{=}C$$ (b,a)
(a) 4.3
(b) 1.3

$$CH_3CH_2ONO$$ (b,a)
(a) 4.78
(b) 1.39

Isopropyl esters

$$(CH_3)_2CHOCOCH_3$$ (b,a)
(a) 4.94
(b) 1.22

$$(CH_3)_2CHOSO_2\text{-}C_6H_4\text{-}CH_3$$ (b,a)
(a) 4.70
(b) 1.25

$$(CH_3)_2CHOCO\text{-}C_6H_5$$ (b,a)
(a) 5.22
(b) 1.37

^{1}H-Chemical Shifts (δ in ppm relative to TMS) and Coupling Constants (J in Hz) in Lactones

β-propiolactone (a,b)
(a) 4.29
(b) 3.56

γ-butyrolactone (a,b,c)
(a) 4.3
(b) 2.1
(c) 2.3

(a) 4.92
(b) 7.63
(c) 6.15

(a) 4.09
(b) } ∼1.6
(c) }
(d) 3.31

(a) 7.77
(b) 6.43
(c) 7.56
(d) 6.38

$J_{ab} = 5.0$ $J_{ac} = 2.4$
$J_{bc} = 6.3$ $J_{bd} = 1.5$
$J_{cd} = 9.4$ $J_{ad} = 1.3$

^{1}H-Chemical Shifts (δ in ppm relative to TMS) and Coupling Constants (J in Hz) in Aliphatic Amides

Chemical Shifts of the NH-Protons

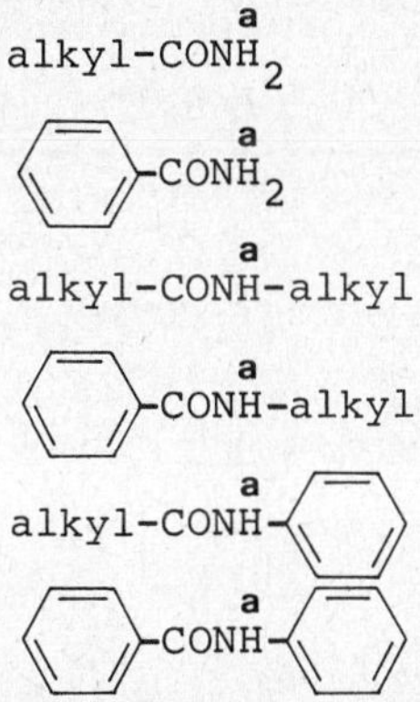

alkyl-CONH$_2$ (a)

◯-CONH$_2$ (a)

$\Big\}$ (a) 5-7

alkyl-CONH-alkyl (a)

◯-CONH-alkyl (a)

$\Big\}$ (a) 6-8.5

alkyl-CONH-◯ (a)

◯-CONH-◯ (a)

$\Big\}$ (a) 7.5-9.5

The signals for the NH-protons are often broad because the ^{14}N-^{1}H coupling is only partly eliminated by the quadrupole relaxation.

Vicinal Coupling H-C-N-H

The HC-NH-CO coupling depends on the conformation analogous to the HC-CH coupling (see p. H20). For free rotation:

$$J_{\underline{H}C-N\underline{H}-CO} = \sim 7$$

This coupling which causes splitting of the C-H signal can be clearly observed even in those cases where the signal for the NH-proton is broad and featureless.

Slow Rotation

The rotation around the CO-N bond is usually so slow that two separate signals are observed for the two conformers. In general the following holds:

for NC$\underline{H}_3$, NCH$_2$C$\underline{H}_3$ and NCH(C$\underline{H}_3$)$_2$ $\delta_{cis\ to\ O} \lessgtr \delta_{trans\ to\ O}$

for NC$\underline{H}$(CH$_3$)$_2$ and NC(C$\underline{H}_3$)$_3$ $\delta_{trans\ to\ O} \lessgtr \delta_{cis\ to\ O}$

for N-C$\underline{H}_2$- $\delta_{cis\ to\ O} \sim \delta_{trans\ to\ O}$

Formamides

In the more stable conformer of monosubstituted formamides the substituent occupies the cis-position relative to the carbonyl oxygen:

$$
\begin{array}{cc}
\underset{\text{H}}{\overset{a}{}} \quad \underset{\text{H}}{\overset{b}{}} & \underset{\text{H}}{\overset{a}{}} \quad \underset{CH_3}{\overset{c}{}}
\end{array}
$$

I (∿90%) II (∿10%)

I:	(a) 8.1	II:	(a) 8.1
	(b) 7.9		(b) 7.9
	(c) 2.74		(c) 2.88

$$\underset{H-CONH-}{\overset{a\qquad b}{}}\bigcirc$$

(a) 8.2–8.7
(b) 7.5–9.5

$$\underset{O=C-N}{\overset{a\qquad\; b}{}}\begin{smallmatrix}CH_3\\ CH_3\end{smallmatrix}$$

(a) 8.02
(b) 2.97
(c) 2.88

$|J|_{ab} = \text{∿}0.3$

$|J|_{ac} = \text{∿}0.7$

In the more stable conformer of disubstituted formamides the larger substituent occupies the trans-position relative to the carbonyl oxygen:

$$\text{H}\quad CH(CH_3)_2 \qquad \longleftarrow\!\!\!\longrightarrow \qquad H \quad CH_3$$

I (∿70%) II (∿30%)

I:	(a) 2.71	II:	(a) 2.83
	(b) 4.12		(b) 4.78
	(c) 1.19		(c) 1.10

Acetamides

In mono-substituted acetamides the only observable conformer has the substituent cis to the carbonyl oxygen:

$$\overset{a\;\;\; b\quad\; c}{CH_3NHCOCH_3}$$

(a) 2.71
(b) 8.1
(c) ∿2.0

$$\overset{a\quad b\;\; c\quad d}{CH_3CH_2NHCOCH_3}$$

(a) 1.12
(b) 3.21
(c) 8.2
(d) ∿2.0

$(CH_3)_2CHNHCOCH_3$ a b c d

(a) 1.13
(b) 4.01
(c) 8.1
(d) ∿2.0

$(CH_3)_3CNHCOCH_3$ a b c

(a) 1.28
(b) 7.3
(c) ∿2.0

CH_3CONH-⟨phenyl⟩   a

(a) ∿2.1

⟨dimethylacetamide structure, a = CH₃, b/c = N-CH₃⟩

(a) 2.08
(b) 2.94
(c) 3.02

⟨N-acetylpyrrolidine structure, a = CH₃CO-N, b, c, d, e⟩

(a) 2.05
(b,e) 3.46
(c,d) 1.97

alkyl-CO-N⟨piperidine, a, b⟩

(a,b) 3.36

In the more stable conformer of a disubstituted acetamide the
larger substituent is cis to the carbonyl oxygen:

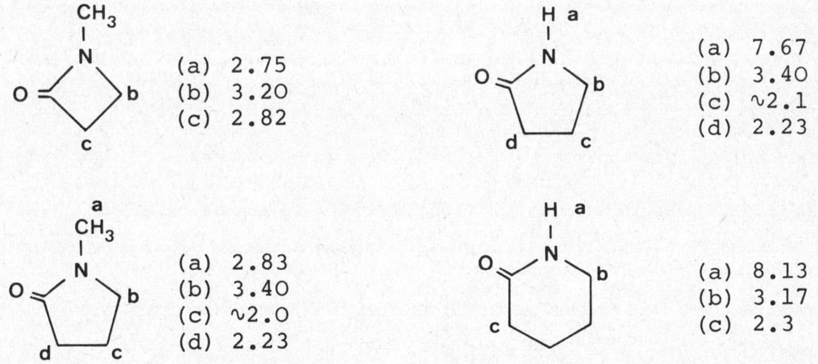

I (∿40%)

I: (a) 2.70
   (b) 3.92
   (c) 1.15

II (∿60%)

II: (a) 2.83
    (b) 4.52
    (c) 1.03

¹H-Chemical Shifts in Lactams (δ in ppm relative to TMS)

⟨β-lactam, N-CH₃ (a), b, c⟩

(a) 2.75
(b) 3.20
(c) 2.82

⟨γ-lactam (2-pyrrolidinone), N-H (a), b, c, d⟩

(a) 7.67
(b) 3.40
(c) ∿2.1
(d) 2.23

⟨N-methyl-2-pyrrolidinone, N-CH₃ (a), b, c, d⟩

(a) 2.83
(b) 3.40
(c) ∿2.0
(d) 2.23

⟨δ-lactam (2-piperidinone), N-H (a), b, c⟩

(a) 8.13
(b) 3.17
(c) 2.3

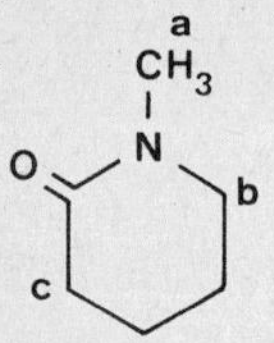

^{1}H-NMR

LACTAMS, IMIDES

	(a) 2.88
	(b) 3.32
	(c) 2.27

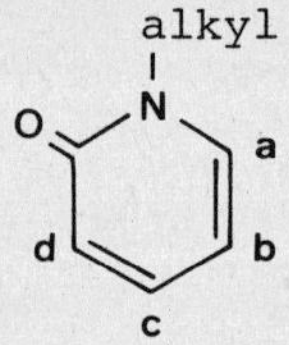

	(a) 7.3
	(b) 6.2
	(c) 7.3
	(d) 6.6

^{1}H-Chemical Shifts in Imides (δ in ppm relative to TMS)

alkyl-CONHCO-alkyl   (a)   8 - 12

The signals are often broadened because the ^{14}N-^{1}H coupling is only partially eliminated by quadrupole relaxation.

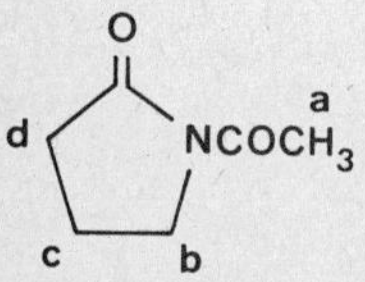

	(a) 2.50
	(b) 3.83
	(c) 2.06
	(d) 2.62

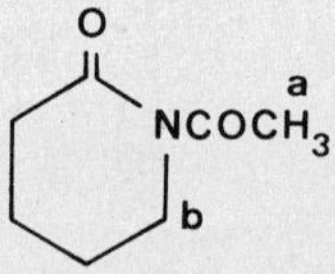

	(a) 2.36
	(b) 3.48

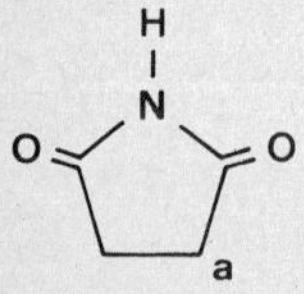

(a) 2.73

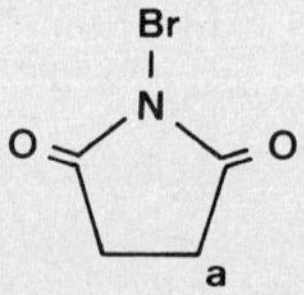

(a) 2.97

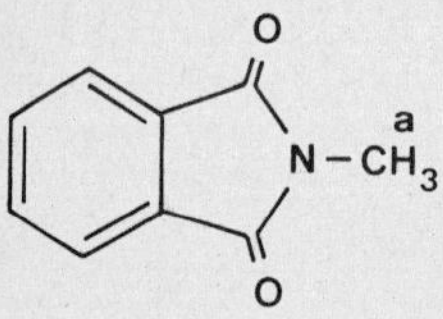

(a) 3.16

II165

^{1}H-Chemical Shifts in Acid Halides and Anhydrides

(δ in ppm relative to TMS)

$\overset{a}{C}H_3COCl$ (a) 2.8

$\overset{a}{C}H_3COBr$ (a) 2.7

$\overset{a}{C}H_3COOCOCH_3$ (a) 2.2

(succinic anhydride)   (a) 3.01

(maleic anhydride)   (a) 7.1

^{1}H-Chemical Shifts in Carbonic Acid Derivatives

(δ in ppm relative to TMS)

$O=C\begin{cases} O\overset{a}{C}H_2\overset{b}{C}H_2\overset{c}{C}H_2\overset{d}{C}H_3 \\ OCH_2CH_2CH_2CH_3 \end{cases}$ (a) 4.13   (b,c) 1.2-1.7   (d) 0.93

(ethylene carbonate)   (a) 4.20

(ethylene trithiocarbonate)   (a) 3.94

$\overset{a}{C}H_3NH\overset{b}{C}ONHCH_3$ (a) 2.7   (b) ∿6

$\overset{a}{C}H_3\overset{b}{C}H_2NHCONHCH_2CH_3$ (a) 1.26   (b) 3.27

$\overset{a}{C}H_3\overset{b}{C}H_2NH\overset{c}{C}ONH_2^{\,d}$ (a) 1.5   (b) 3.0   (c) 5.5   (d) 5

$\overset{a}{C}H_3NH\overset{b}{C}OO\overset{c}{C}H_2\overset{d}{C}H_3$ (a) 2.78   (b) 5.16   (c) 4.14   (d) 1.23

^{1}H-NMR

OXIMES, HYDRAZONES

^{1}H-Chemical Shifts in Oximes, Imines, Hydrazones and Azines

(δ in ppm relative to TMS)

| alkyl—CH(a)=N—OH(b) | (a) 6.8–7.9
 (b) 7–10 | $\delta_{a\ syn} > \delta_{a\ anti}$ |

| aryl—CH(a)=N—OH(b) | (a) 7.2–8.6
 (b) 7–10 | $\delta_{a\ syn} > \delta_{a\ anti}$ |

| alkyl—CH(a)=N—NH—aryl | (a) 6.1–7.7 | $\delta_{a\ syn} > \delta_{a\ anti}$ |

| CH_3(b)—C(H,a)=N—OH(c) | (a) 6.8
 (b) 1.9
 (c) 9 | | CH_3(b)—C(H,a)=N—OH(c) | (a) 7.4
 (b) 1.9
 (c) 9 |

In aldoximes and ketoximes $\Delta\delta = \delta_{syn} - \delta_{anti}$ depends on the dihedral angle θ (H–C–C=N):

θ	$\Delta\delta$ (ppm)
$0°$	1
$60°$	0
$115°$	−0.3

$$CH_3-N=CH-\text{(phenyl)}$$

    (a) 3.4
    (b) 8.40

$$CH_3-C(H)=N-N=C(H)-CH_3$$

    (a) 2.03
    (b) 7.89

$$CH_3-C(CH_3)=N-N=C(CH_3)-CH_3$$

    (a) 2.00
    (b) 1.83

---

^{1}H-Chemical Shifts (δ in ppm relative to TMS) and Coupling Constants

(J in Hz) in Nitriles, Isonitriles, Cyanates and Isocyanates

For thiocyanates and isothiocyanates see p. H115.

CH_3CN	(a) 1.98	$(CH_3)_2CHCN$	(a) 2.67 (b) 1.35
CH_3CH_2CN	(a) 2.35 (b) 1.31	$(CH_3)_3CCN$	(a) 1.37
CH_3NC	(a) 2.85	$(CH_3)_2CHNC$	(a) 4.83 (b) 1.45

In isonitriles the quadrupole relaxation of the nitrogen nucleus
is so small that the ^{14}N-^{1}H couplings become observable:

$$-\overset{\beta}{CH_2}-\overset{\alpha}{CH_2}-{}^{14}NC$$

$$|J|_{H_\alpha N} = 1.8\text{-}2.8$$
$$|J|_{H_\beta N} = 2.5\text{-}3.5$$

CH_3CH_2OCN (a) 4.54
    (b) 1.45

CH_3CH_2NCO (a) 3.37
    (b) 1.20

^{1}H-Chemical Shifts (δ in ppm relative to TMS) and Coupling Constants
(J in Hz) in Saturated Alicyclics

For unsaturated alicyclics see p. H230 - H240.

0.22   geminal coupling: J_{gem} = -3...-9

vicinal couplings: J_{cis} = 7...13

J_{trans} = 4...9.5

always $J_{cis} > J_{trans}$

1.96   geminal coupling: J_{gem} = -11...-17

vicinal couplings: J_{cis} = 4...12

J_{trans} = 2...10

"long-range" coupling: J = 0.5...2.0

1.51   geminal coupling: J_{gem} = -8...-18

vicinal couplings: J_{cis} = 5...10

J_{trans} = 5...10

1.44   geminal coupling: J_{gem} = -11...-14

(for -100°C: vicinal couplings: $J_{ax,ax}$ = 8...13

H_{ax}: 1.1, $J_{aq,ax}$ = 2...6

H_{eq}: 1.6) $J_{eq,eq}$ = 2...5

generally $J_{eq,ax} \sim J_{eq,eq} + 1$

H_a
H_b

(a) 1.8
(b) 1.9

In condensed alicyclics "long-range" couplings (coupling over 4 or more bonds) are often observed. They are particularly large if the bonds between the two protons are w-shaped:

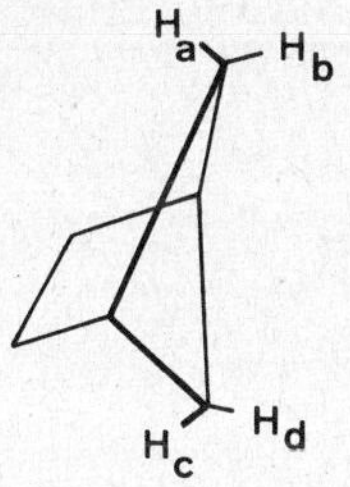

$$J_{ac} = \sim 7$$
$$J_{ad} = J_{bd} = \sim 0$$

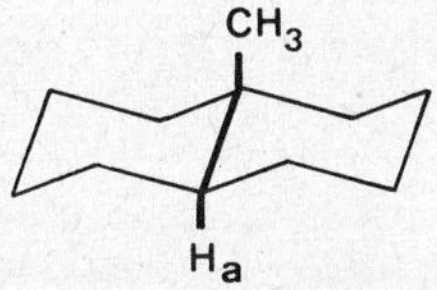

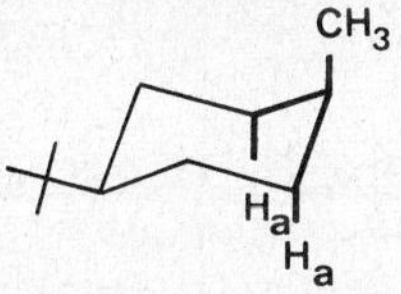

the signal of the methyl group is significantly broadened due to the "long-range" coupling with H_a

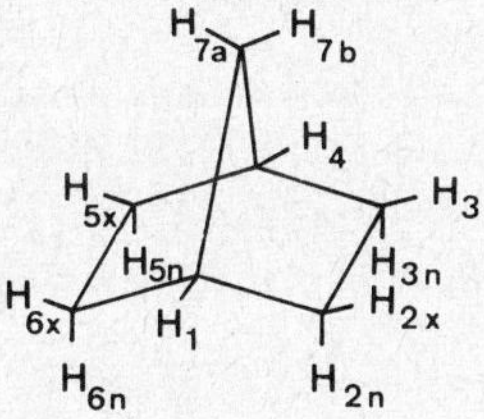

H_1	2.19	$|J|_{1,2n} = 0-2$	$J_{7b,2n} = \sim 0$
H_{2n}	1.18	$|J|_{1,2x} = 3-4$	$J_{7a,2x} = \sim 0$
H_{2x}	1.46	$|J|_{2n,2x} = 2.5-5$	$J_{7b,2x} = \sim 0$
H_7	1.18	$|J|_{2n,3n} = 6-7$	$J_{2x,6n} = \sim 0$
		$|J|_{2x,3x} = 9-10$	$J_{2n,6n} = \sim 0$
		$|J|_{1,4} = 1-1.5$	
		$|J|_{2x,6x} = 1-1.5$	
		$|J|_{7a,2n} = 3-4$	

^{1}H-NMR

SUBSTITUTED CYCLOHEXANES

^{1}H-Chemical Shifts of Axial and Equatorial Protons in Substituted Cyclohexanes[1] (δ in ppm relative to TMS)

Substituent X	$\delta_{H_{ax}}$	$\delta_{H_{eq}}$
-D	1.12	1.60
-CH$_3$	1.27	1.93
-phenyl	2.47	2.98
-Cl	3.63	4.34
-Br	3.81	4.62
-I	3.98	4.72
-OH	3.38	3.89
-OCOCH$_3$	4.46	4.98
-NH$_2$	2.52	3.15
-NHCH$_3$	2.08	2.70
-NO$_2$	4.23	4.43
-SH	2.57	3.43

[1] These values were determined by measurements at low temperatures or on 4-tert.-butyl-substituted cycloalkanes.

H195

Effect of the Substituent on the Chemical Shift of Protons
in Cyclohexane (in ppm, a negative sign indicates shielding
by the substituent)

Substituent X, position	H_{1} ax or eq	H_{2} ax	H_{2} eq	H_{3} ax	H_{3} eq	H_{4} ax	H_{4} eq
-CH$_3$ eq	0.15	-0.31	-0.03	0.03	0	-0.06	-0.02
ax	0.33	0.25	-0.20	0.27	-0.26	-0.06	-0.02
-Cl eq	2.51						
ax	2.74			0.6			
-OH eq	2.26	-0.03	0.18	0.07	0.01	0.01	0.06
ax	2.29	0.23	-0.02	0.46	-0.27	0.02	-0.08
-OCOCH$_3$ eq	3.34						
ax	3.38	0.35	∿0.7				
-NO$_2$ eq	3.11	1.1	0.3				
ax	2.83	0.5	1.0				
-SH eq	1.45	-0.4	-0.3				
ax	1.83		-0.1	0.8			

^{1}H-Chemical Shifts (δ in ppm relative to TMS) and Coupling Constants
(J in Hz) in Alkenes

Ethylene

5.28

geminal coupling: $J_{gem} = 2.5$

vicinal couplings: $J_{cis} = 11.6$

$J_{trans} = 19.1$

The coupling constants depend strongly on the electronegativity of
the substituents. They decrease with increasing electronegativity
of the substituents and their number (the contribution of the geminal
coupling constant may, however, increase because J_{gem} is often negative in substituted ethylenes). J_{trans} is always > J_{cis}.

R	J_{ab}	J_{ac}	J_{bc}
-Li	19.3	23.9	7.1
-CH$_3$	10.0	16.8	2.1
-F	4.7	12.7	-3.2

Coupling Over More than Three Bonds ("Long-Range" Coupling)

If there is a double bond between two coupling nuclei, coupling
across four (allylic coupling) or five (homoallylic coupling)
bonds may be observed.

Allylic Couplings

cisoid: $J_{ab} = -3.....+2$

transoid: $J_{ac} = -3.5...+2.5$

the magnitude of the coupling constant
depends on the conformation:

ϕ	J_{ab}	J_{ac}
0°	-3.0	-3.5
90°	+1.8	+2.2
180°	-3.0	-3.5
270°	0	0.8

The frequently used rough rule that $|J|_{cisoid} > |J|_{transoid}$ does not hold generally. It most often holds in acyclic systems.

<u>Homoallylic coupling</u>

cisoid: $|J|_{ab} = 0...3$
transoid: $|J|_{ac} = 0...3$

generally: $J_{H-C=C-CH_3} \sim -J_{CH_3-C=C-CH_3}$

In acyclic systems the relationship $|J|_{cisoid} < |J|_{transoid}$ generally holds.

Large homoallylic couplings are generally observed in cyclic systems:

$J_{ab} = 5...11$

$J_{ab} = 0...7$

X : CH, N   R : any substituent   X : O, NH

<u>Butadiene</u>

(a) 5.16 $J_{ab} = 1.8$ $J_{bc} = 10.2$
              $J_{ac} = 17.1$ $J_{bd} = -0.9$
(b) 5.06 $J_{ad} = -0.8$ $J_{be} = 1.3$
              $J_{ae} = 0.6$ $J_{cd} = 10.4$
(c) 6.27 $J_{af} = 0.7$

<u>Allene</u>

4.67 $J_{ab} = -9$
          $J_{ac} = -6$

$|J|_{ab} = 3.0$

^{1}H-NMR

ALKENES, ADDITIVITY RULE

The Chemical Shift of Protons at a Double Bond

(δ in ppm relative to TMS)

$$\delta_{C=CH} = 5.25 + Z_{gem} + Z_{cis} + Z_{trans}$$

	Substituent R	Z_{gem}	Z_{cis}	Z_{trans}
C	-H	0	0	0
	-alkyl	0.45	-0.22	-0.28
	-alkyl ring[1]	0.69	-0.25	-0.28
	-CH$_2$-aromatic	1.05	-0.29	-0.32
	-CH$_2$X, X: F, Cl, Br	0.70	0.11	-0.04
	-CHF$_2$	0.66	0.32	0.21
	-CF$_3$	0.66	0.61	0.32
	-CH$_2$O	0.64	-0.01	-0.02
	-CH$_2$N	0.58	-0.10	-0.08
	-CH$_2$S	0.71	-0.13	-0.22
	-CH$_2$CO, CH$_2$CN	0.69	-0.08	-0.06
	-C=C isolated	1.00	-0.09	-0.23
	-C=C conjugated[2]	1.24	0.02	-0.05
	-C≡C	0.47	0.38	0.12
	-aromatic free rotation	1.38	0.36	-0.07
	-aromatic fixed[3]	1.60	–	-0.05
	-aromatic o-substituted	1.65	0.19	0.09
H A L	-F	1.54	-0.40	-1.02
	-Cl	1.08	0.18	0.13
	-Br	1.07	0.45	0.55
	-I	1.14	0.81	0.88

[1] The increment for "alkyl ring" is to be used if the substituent and the double bond are part of a cyclic structure.

[2] The increment "C=C conjugated" is to be used if either the double bond or the C=C substituent is conjugated to other substituents.

[3] The increment "aromatic, fixed" is to be used if the double bond conjugated to an aromatic ring is part of a fused ring (such as in 1,2-dihydronaphthalene).

Substituent R	Z_{gem}	Z_{cis}	Z_{trans}
O -OR, R aliphatic	1.22	-1.07	-1.21
-OR, R unsaturated	1.21	-0.60	-1.00
-OCOR	2.11	-0.35	-0.64
$-NH_2$	0.80	-1.26	-1.21
-NHR, R aliphatic	0.80	-1.26	-1.21
$-NR_2$, R aliphatic	0.80	-1.26	-1.21
N -NHR, R unsaturated	1.17	-0.53	-0.99
-NRR', R unsaturated, R' any substituent	1.17	-0.53	-0.99
-NCOR	2.08	-0.57	-0.72
-N=N-phenyl	2.39	1.11	0.67
$-NO_2$	1.87	1.30	0.62
-SR	1.11	-0.29	-0.13
-SOR	1.27	0.67	0.41
S $-SO_2R$	1.55	1.16	0.93
-SCOR	1.41	0.06	0.02
-SCN	0.80	1.17	1.11
$-SF_5$	1.68	0.61	0.49
-CHO	1.02	0.95	1.17
-CO isolated	1.10	1.12	0.87
-CO conjugated [1]	1.06	0.91	0.74
-COOH isolated	0.97	1.41	0.71
-COOH conjugated [1]	0.80	0.98	0.32
-COOR isolated	0.80	1.18	0.55
-COOR conjugated [1]	0.78	1.01	0.46
$-CONR_2$	1.37	0.98	0.46
-COCl	1.11	1.46	1.01
-CN	0.27	0.75	0.55
$-PO(OCH_2CH_3)_2$	0.66	0.88	0.67
$-OPO(OCH_2CH_3)_2$	1.33	-0.34	-0.66

[1] The increment "conjugated" is to be used if either the double bond or the substituent is conjugated to additional substituents.

^{1}H-NMR

ALKENES, ALKYNES

^{1}H-Chemical Shifts in Substituted Isobutenes

(δ in ppm relative to TMS)

R	δ_a	δ_b	δ_c
-H		1.70	1.70
-C(CH$_3$)$_3$	5.13	1.68	1.62
-C≡CH	5.17	1.80	1.88
-Br	5.78	1.75	1.75
-OCOCH$_3$	6.79	1.65	1.65
-CHO	5.63	1.91	2.11
-COCH$_3$	5.97	1.86	2.06
-COOCH$_3$	5.62	1.84	2.12
-COCl	6.01	1.97	2.12

^{1}H-Chemical Shifts (δ in ppm relative to TMS) and Coupling Constants (J in Hz) in Acetylene Derivatives

H-C≡C-H	1.80	CH$_3$-C≡C-H	(a)	1.80
H-C≡C-alkyl	1.7-1.9		(b)	1.80
H-C≡C-C=C	2.6-3.1	CH$_3$CH$_2$-C≡CH	(a)	2.16
H-C≡C-C≡C	1.7-2.4		(b)	1.15
		(CH$_3$)$_2$CH-C≡CH	(a)	2.59
H-C≡C-phenyl	2.7-3.4		(b)	1.15
		phenyl-SO$_3$CH$_2$-C≡CH	(a)	∿4.7
H-C≡C-O-alkyl	1.3		(b)	2.55
H-C≡C-CO	2.1-3.3	CH$_3$CONHCH$_2$-C≡CH	(a)	4.06
			(b)	2.25

CH$_3$-C≡C-H |J| = 2.9

CH$_3$-C≡C-CH$_3$ |J| = 2.7

H-C≡C-C≡C-H |J| = 2.2

H225

<u>^{1}H-Chemical Shifts (δ in ppm relative to TMS) and Coupling Constants</u>

<u>(J in Hz) in Unsaturated Alicyclics</u>

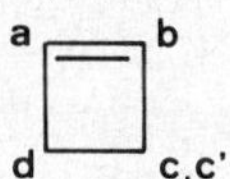

(a,b) 7.01   in derivatives: J_{ab} = 0.5-1.5

(c)   0.92 J_{bc} = 1.8

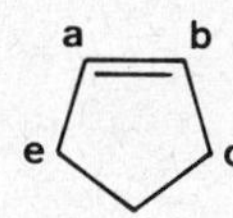

(a,b) 5.95   geminal coupling: $J_{cc'}$ = -12.0

(c,d) 2.57   vicinal couplings: J_{ab} = 2.7 (in deriva-
                                            tives: 2.5-4.0)

J_{bc} = -0.8

$J_{cd(cis)}$ = 4.4

$J_{cd(trans)}$ = 1.7

"long-range" coupling: J_{ac} = 1.6

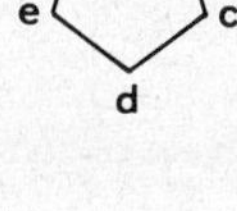

(a,b) 5.60   in derivatives: J_{ab} = 5.0-7.0

(c,e) 2.28 J_{bc} = 0.5

(d)   1.90

(a,e) 6.5   vicinal couplings: J_{ab} = 5.1

(b,d) 6.4 J_{bc} = 1.2

(c)   2.90 J_{ae} = 1.9

               "long-range" couplings: J_{ac} = -1.3

J_{ad} = 1.1

J_{bd} = 1.9

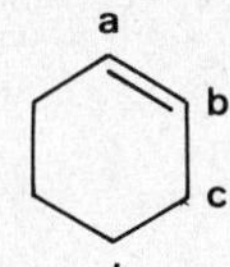

(a,b) 5.59   in derivatives: J_{ab} = 8.5-11.0

(c)   1.96 J_{bc} = 1.5

(d)   1.65

^{1}H-NMR

(a,d) 5.8   vicinal couplings: J_{ab} = 9.4

(b,c) 5.9 J_{bc} = 5.1

(e,f) 2.15   "long-range" couplings: J_{ac} = 1.1

                                              J_{ad} = 0.9

(a,b) 5.71   in derivatives: J_{ab} = 9.0-12.5

(c)   2.11 J_{bc} = 3.7

(a,f) 5.26   geminal coupling: $J_{gg'}$ = -13.0

(b,e) 6.09   vicinal couplings: J_{ab} = 8.9

(c,d) 6.50 J_{bc} = 5.5

(g)   2.22 J_{cd} = 11.2

                                           J_{ag} = 6.7

"long-range" couplings: J_{ac} + J_{ad} = 1.5

                                        J_{ae} = J_{af} = 0

                                          J_{bd} = 0.8

                                          J_{be} = -0.6

                                          J_{bf} = 0

(a,b) 5.56   in derivatives: J_{ab} = 10-13

(c)   2.11 J_{bc} = 5.3

(d,e) 1.5

^{1}H-Chemical Shifts (δ in ppm relative to TMS) and Coupling Constants
(J in Hz) in Alicyclic Rings Condensed with Aromatic Systems

(See also: cyclic ketones, p. H130.)

(a,b)	~7.2	
(c)	2.91	
(d)	2.04	

(a)	3.33	$J_{ab} = 2.0$	$J_{bc} = 5.8$
(b)	6.50	$J_{ac} = 2.0$	$J_{cd} = 0.7$
(c)	6.82		

(a)	7.84	$J_{ab} = 7.5$	$J_{bc} = 6.5$
(b)	7.38	$J_{ac} = 1.6$	$J_{bd} = 1.6$
(c)	7.28	$J_{ad} = 0.2$	$J_{cd} = 7.2$
(d)	7.55		
(e)	3.87		

(a)	6.93	
(b)	7.01	
(c)	2.85	
(d)	1.60	

(a)	3.91
(b)	7.31
(c)	7.19

(a)	2.86

For acenaphthene and acenaphthylene see p. H250.

<u>^{1}H-Chemical Shifts (δ in ppm relative to TMS) and Coupling Constants (J in Hz) in Aromatic Hydrocarbons</u>

(Substituted benzenes: see p. H255, H260.)

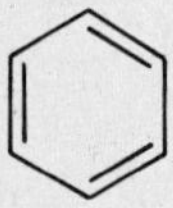

7.26   in derivatives: J_{ortho} = 6.0-9.0
J_{meta} = 1.0-3.0
J_{para} = 0-1.0

"Long-range" couplings with the protons of substituents:

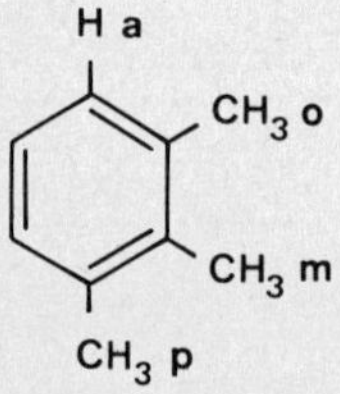

J_{ao} = -0.6...-0.9
J_{am} = 0...+0.3
J_{ap} = ∿-0.6

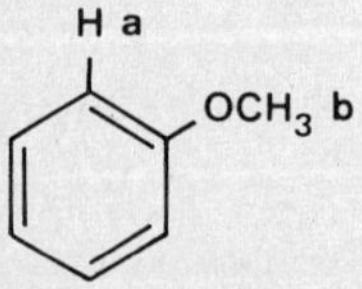

J_{ab} = 0-0.8

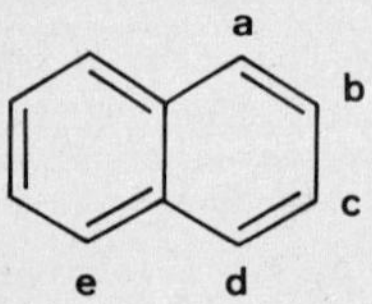

(a) 7.66 J_{ab} = 8.5 (in derivatives: 8-9)
(b) 7.30 J_{bc} = 7.5 (in derivatives: 5-7)
J_{ac} = 1.4 (in derivatives: 1-2)
J_{ad} = 0.7 (in derivatives: ∿1)
J_{ae} in derivatives: ∿1

(a) 7.15 $J_{ab} = 0$ $J_{bc} = 7$
(b) 7.90 $J_{ac} = 0$ $J_{bd} = 0.6$
(c) 7.58 $J_{ad} = 0$ $J_{cd} = 8$
(d) 7.79

(a) 3.34 $J_{bc} = 6.7$ $J_{ab} = 1.5$
(b) 7.11 $J_{cd} = 8.1$ $J_{ad} = 0.5$
(c) 7.31 $J_{bd} = 1.2$
(d) 7.46

(a) 8.31 $J_{bc} = 8.4$ $J_{be} = 0.5$
(b) 7.91 $J_{bd} = 1.5$ $J_{cd} = 6.0$
(c) 7.39

(a) 8.93 $J_{ab} = 8.4$ $J_{bc} = 7.2$
(b) 7.88 $J_{ac} = 1.2$ $J_{bd} = 1.3$
(c) 7.82 $J_{ad} = 0.7$ $J_{cd} = 8.1$
(d) 8.12
(e) 7.71

^{1}H-NMR

Effect of a Substituent on the Chemical Shift of the Ring-Protons in Benzene (δ in ppm relative to TMS)

$$\delta_{H_i} = 7.26 + Z_i$$

	Substituent X	Z_2	Z_3	Z_4
	-H	0	0	0
	$-CH_3$	-0.20	-0.12	-0.22
	$-CH_2CH_3$	-0.14	-0.06	-0.17
	$-CH(CH_3)_2$	-0.13	-0.08	-0.18
	$-C(CH_3)_3$	0.02	-0.08	-0.21
	$-CH_2Cl$	0.00	0.00	0.00
	$-CF_3$	0.32	0.14	0.20
C	$-CCl_3$	0.64	0.13	0.10
	$-CH_2OH$	-0.07	-0.07	-0.07
	$-CH=CH_2$	0.06	-0.03	-0.10
	-CH=CH-phenyl	0.15	-0.01	-0.16
	$-C\equiv CH$	0.15	-0.02	-0.01
	$-C\equiv C$-phenyl	0.19	0.02	0.00
	-phenyl	0.37	0.20	0.10
	-F	-0.26	0.00	-0.20
H	-Cl	0.03	-0.02	-0.09
A	-Br	0.18	-0.08	-0.04
L	-I	0.39	-0.21	0.00
	-OH	-0.56	-0.12	-0.45
	$-OCH_3$	-0.48	-0.09	-0.44
	$-OCH_2CH_3$	-0.46	-0.10	-0.43
O	-O-phenyl	-0.29	-0.05	-0.23
	$-OCOCH_3$	-0.25	0.03	-0.13
	-OCO-phenyl	-0.09	0.09	-0.08
	$-OSO_2CH_3$	-0.05	0.07	-0.01

	Substituent X	z_2	z_3	z_4
N	$-NH_2$	-0.75	-0.25	-0.65
	$-NHCH_3$	-0.80	-0.22	-0.68
	$-N(CH_3)_2$	-0.66	-0.18	-0.67
	$-N^+(CH_3)_3$ I^-	0.69	0.36	0.31
	$-NHCOCH_3$	0.12	-0.07	-0.28
	$-N(CH_3)COCH_3$	-0.16	0.05	-0.02
	$-NHNH_2$	-0.60	-0.08	-0.55
	$-N=N-phenyl$	0.67	0.20	0.20
	$-NO$	0.58	0.31	0.37
	$-NO_2$	0.95	0.26	0.38
S	$-SH$	-0.08	-0.16	-0.22
	$-SCH_3$	-0.08	-0.10	-0.24
	$-S-phenyl$	0.06	-0.09	-0.15
	$-SO_3CH_3$	0.60	0.26	0.33
	$-SO_2Cl$	0.76	0.35	0.45
	$-CHO$	0.56	0.22	0.29
	$-COCH_3$	0.62	0.14	0.21
	$-COCH_2CH_3$	0.63	0.13	0.20
	$-COC(CH_3)_3$	0.44	0.05	0.05
$\overset{O}{\underset{\diagdown}{\underset{C}{\parallel}}}$	$-CO-phenyl$	0.47	0.13	0.22
	$-COOH$	0.85	0.18	0.27
	$-COOCH_3$	0.71	0.11	0.21
	$-COOCH(CH_3)_2$	0.70	0.09	0.19
	$-COO-phenyl$	0.90	0.17	0.27
	$-CONH_2$	0.61	0.10	0.17
	$-COCl$	0.84	0.22	0.36
	$-COBr$	0.80	0.21	0.37
	$-CH=N-phenyl$	~0.6	~0.2	~0.2
	$-CN$	0.36	0.18	0.28
	$-Si(CH_3)_3$	0.22	-0.02	-0.02
	$-PO(OCH_3)_2$	0.48	0.16	0.24

^{1}H-NMR

5-MEMBERED HETEROAROMATIC RINGS

^{1}H-Chemical Shifts (δ in ppm relative to TMS) and Coupling Constants (J in Hz) in Non-Condensed Heteroaromatics

For non-aromatic heterocycles see: cyclic ethers, amines and thio-ethers.

(a,d) 7.38	J_{ab} = 1.8	J_{ad} = 1.5
(b,c) 6.30	J_{ac} = 0.9	J_{bc} = 3.4

Substituted furans: see p. H285, H290.

(a) 7-12 (very solvent-dependent, broad)		
(b,e) 6.62	J_{ab} = 2.6	J_{bd} = 1.3
(c,d) 6.05	J_{ac} = 2.3	J_{be} = 2.1
	J_{bc} = 2.6	J_{cd} = 3.5

Substituted pyrroles: see p. H295, H300.

(a,d) 7.20	J_{ab} = 4.8	J_{ad} = 2.8
(b,c) 6.96	J_{ac} = 1.0	J_{bc} = 3.5

Substituted thiophenes: see p. H305, H310.

(a,d) 7.70	J_{ab} = 5.4	J_{ad} = 2.5
(b,c) 7.12	J_{ac} = 1.1	J_{bc} = 3.6

imidazole (H d)

(a,b)	7.13
(c)	7.70
(d)	13.4

in derivatives: $J_{ab} = 1\text{-}2$

$J_{ac} = 1\text{-}2$

$J_{bc} = 0.5\text{-}1.5$

imidazolium cation

(a,b)	7.6
(c)	∿12
(d)	8.6

$J_{ac} + J_{bc} = 4.4$

$J_{cd} = 2.4$

$J_{ad} = 1.4$

pyrazole (H d)

(a,c)	7.55
(b)	6.25
(d)	13.7

$J_{ab} = 2.1$ in derivatives: $J_{ab} = 2\text{-}3$

$J_{bc} = 1\text{-}2$

$J_{ac} = 0.5\text{-}1$

oxazole

(a)	7.69
(b)	7.09
(c)	7.95

$J_{ab} = 0.8$

$J_{ac} = 0.5$

thiazole

(a)	7.41
(b)	7.98
(c)	8.88

$J_{ab} = 3.2$

$J_{ac} = 1.9$

$J_{bc} = 0$

isothiazole

(a)	8.72
(b)	7.26
(c)	8.56

$J_{ab} = 4.7$

$J_{ac} < 0.4$

$J_{bc} = 1.7$

^{1}H-NMR

HETEROAROMATICS

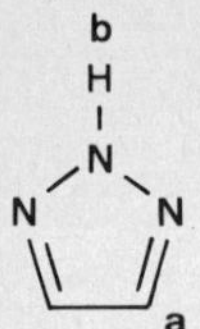

(a) 7.75
(b) 12

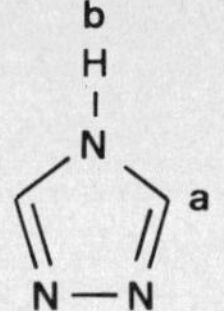

(a) 8.27
(b) 13.5

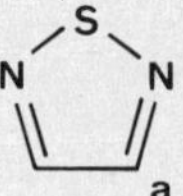

(a) 8.19

(a) 8.58

(a,e)	8.60	J_{ab} = 5.5
(b,d)	7.25	J_{ac} = 1.9
(c)	7.64	J_{ad} = 0.9
		J_{ae} = 0.4
		J_{bc} = 7.6
		J_{bd} = 1.6

in derivatives: J_{ab} = 4-6

J_{ac} = 0-2.5

J_{ad} = 0-2.5

J_{ae} = 0-0.6

J_{bc} = 7-9

J_{bd} = 0.5-2

Substituted pyridines: see p. H315, H320.

(a,e) 8.10
(b,d) 7.28
(c)   7.08

(a,e)	9.23	J_{ab} = 6.0	J_{ae} = 1.0
(b,d)	8.50	J_{ac} = 1.5	J_{bc} = 8.0
(c)	9.04	J_{ad} = 0.8	J_{bd} = 1.4

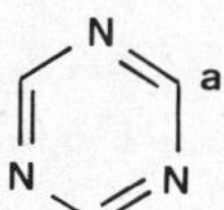

(a,d) 9.24	J_{ab} = 4.9
(b,c) 7.55	J_{ac} = 2.0
	J_{ad} = 3.5
	J_{bc} = 8.4

(a) 8.26	J_{ab} = 6.5
(b) 7.83	J_{ac} = 1.0
(c) 7.22	J_{ad} = 1.0
(d) 8.54	J_{bc} = 8.0
	J_{bd} = 2.5

(a,c) 8.78	J_{ab} = 5.0
(b) 7.36	J_{ac} = 2.5
(d) 9.26	J_{ad} = 1.5
	J_{bd} = 0

(a) 8.43	J_{ab} = 6.8	J_{bc} = 4.9
(b) 7.34	J_{ac} = 1.6	J_{bd} = 1.0
(c) 8.24	J_{ad} = 2.0	J_{cd} = 0
(d) 8.98		

(a) 8.63	J_{ab} = 1.8
	J_{ac} = 1.8
	J_{ad} = 0.5

| (a) 9.25 |

| (a) 9.48 |
| (b) 8.84 |
| (c) 9.88 |

Effect of a Substituent on the Chemical Shift of the Ring-Protons
in Monosubstituted Furan (δ in ppm relative to TMS)

(For couplings see p. H265.)

$$\delta_{H_2} = 7.38 + Z_{i2}$$

$$\delta_{H_3} = 6.30 + Z_{i3}$$

$$\delta_{H_4} = 6.30 + Z_{i4}$$

$$\delta_{H_5} = 7.38 + Z_{i5}$$

2- or 5-Substituent (i = 2 or 5)	$Z_{23} = Z_{54}$	$Z_{24} = Z_{53}$	$Z_{25} = Z_{52}$
–H	0	0	0
–CH$_3$	–0.42	–0.12	–0.17
–CH$_2$OH	–0.11	–0.05	–0.08
–CH$_2$NH$_2$	–0.24	–0.06	–0.10
–CH=CHCHO	0.70	0.35	0.42
–Br	–0.02	0.03	–0.01
–I	0.12	–0.13	–0.01
–OCH$_3$	–1.34	–0.23	–0.68
–NO$_2$	1.21	0.55	0.51
–SCH$_3$	–0.12	–0.06	–0.09
–SCN	0.40	0.06	0.10
–CHO	0.93	0.31	0.34
–COCH$_3$	0.81	0.23	0.19
–COCF$_3$	1.34	0.50	0.64
–COOH	0.94	0.33	0.41
–COOCH$_3$	0.85	0.22	0.25
–COCl	1.20	0.39	0.48
–CN	0.85	0.32	0.28

3- or 4-Substituent (i = 3 or 4)	$Z_{32} = Z_{45}$	$Z_{34} = Z_{43}$	$Z_{35} = Z_{42}$
-H	0	0	0
$-CH_3$	-0.27	-0.17	-0.15
-I	-0.13	0.04	-0.22
$-OCH_3$	-0.46	-0.28	-0.37
$-SCH_3$	-0.18	-0.05	-0.15
-SCN	0.19	0.19	0.03
-CHO	0.48	0.37	-0.07
$-COCH_3$	0.46	0.36	-0.12
-COOH	0.89	0.54	0.36
$-COOCH_3$	0.45	0.33	-0.14
-CN	0.45	0.22	-0.02

^{1}H-NMR

PYRROLE, SUBSTITUENT EFFECTS

Effect of a Substituent on the Chemical Shift of the Ring-Protons
in Monosubstituted Pyrrole (δ in ppm relative to TMS)

(For couplings see p. H265.)

$$\delta_{H_1} = 7\text{-}12 \quad \text{(strongly solvent-dependent, generally broad)}$$

$$\delta_{H_2} = 6.62 + z_{i2}$$

$$\delta_{H_3} = 6.05 + z_{i3}$$

$$\delta_{H_4} = 6.05 + z_{i4}$$

$$\delta_{H_5} = 6.62 + z_{i5}$$

1-Substituent ($i = 1$)	$z_{12} = z_{15}$	$z_{13} = z_{14}$
$-H$	0	0
$-CH_3$	-0.25	-0.13
$-CH_2CH_3$	-0.16	-0.12
$-CH_2-$phenyl	-0.12	-0.04
$-$phenyl	0.33	0.14
$-COCH_3$	0.56	0.12

2- or 5-Substituent (i = 2 or 5)	$Z_{23} = Z_{54}$	$Z_{24} = Z_{53}$	$Z_{25} = Z_{52}$
-H	0	0	0
$-CH_3$	-0.33	-0.16	-0.26
$-NO_2$	1.06	0.24	0.43
$-SCH_3$	0.18	0.05	0.10
-SCN	0.48	0.10	0.28
-CHO	0.93	0.27	0.61
$-COCH_3$	0.78	0.10	0.44
$-COOCH_3$	0.79	0.13	0.29
-CN	0.83	0.23	0.51

3- or 4-Substituent (i = 3 or 4)	$Z_{32} = Z_{45}$	$Z_{34} = Z_{43}$	$Z_{35} = Z_{42}$
-H	0	0	0
$-CH_3$	-0.34	-0.20	-0.20
$-NO_2$	1.04	0.70	0.13
$-COCH_3$	0.79	0.63	0.15
$-COOCH_3$	0.90	0.73	0.16

^{1}H-NMR

Effect of a Substituent on the Chemical Shift of the Ring-Protons in Monosubstituted Thiophene (δ in ppm relative to TMS)

(For couplings see p. H265.)

$$\delta_{H_2} = 7.20 + Z_{i2}$$

$$\delta_{H_3} = 6.96 + Z_{i3}$$

$$\delta_{H_4} = 6.96 + Z_{i4}$$

$$\delta_{H_5} = 7.20 + Z_{i5}$$

2- or 5-Substituent (i = 2 or 5)	$Z_{23} = Z_{54}$	$Z_{24} = Z_{53}$	$Z_{25} = Z_{52}$
C -H	0	0	0
-CH$_3$	-0.36	-0.24	-0.29
-C≡CH	0.15	-0.16	-0.12
HAL -Cl	-0.25	-0.22	-0.22
-Br	-0.05	-0.27	-0.11
-I	0.13	-0.33	0.01
O -OH [1]	-0.72	0.59	-3.10
-OCH$_3$	-0.94	-0.43	-0.82
N -NH$_2$	-0.95	-0.45	-0.85
-NO$_2$	0.82	-0.03	0.30
-SH	0.00	-0.20	-0.07
-SCH$_3$	-0.03	-0.18	-0.05
S -SCN	0.30	-0.05	0.28
-SO$_2$CH$_3$	1.03	0.20	0.79
-SO$_2$Cl	0.73	0.06	0.45
-CHO	0.65	0.10	0.45
-COCH$_3$	0.57	0.00	0.28
-COOH	0.80	0.08	0.40
-COOCH$_3$	0.70	-0.05	0.20
-COCl	0.88	0.06	0.44
-CN	0.47	0.00	0.28

[1] present as the keto-form

H305

3- or 4-Substituent (i = 3 or 4)	$z_{32} = z_{45}$	$z_{34} = z_{43}$	$z_{35} = z_{42}$
C -H	0	0	0
-CH$_3$	-0.45	-0.22	-0.14
-Cl	-0.22	-0.11	-0.03
HAL -Br	-0.12	-0.08	-0.10
-I	0.06	0.00	-0.19
O -OCH$_3$	-1.10	-0.38	-0.20
N -NH$_2$	-1.25	-0.53	-0.25
-NO$_2$	0.95	0.60	0.03
-SH	-0.22	-0.20	-0.10
S -SCH$_3$	-0.33	-0.10	-0.03
-SCN	0.25	0.05	0.05
-SO$_2$CH$_3$	0.96	0.48	0.46
-CHO	0.79	0.45	0.03
-COCH$_3$	0.68	0.47	-0.02
-COOH	0.99	0.48	0.24
-COOCH$_3$	0.78	0.47	-0.05
-COCl	1.05	0.50	0.03
-CN	0.63	0.20	0.15

^{1}H-NMR

Effect of a Substituent on the Chemical Shift of the Ring-Protons in Monosubstituted Pyridines

(δ in ppm relative to TMS, solvent: dimethyl sulfoxide)

(For couplings and chemical shifts in $CDCl_3$ as the solvent see p. H275.)

$$\delta_{H_2} = 8.59 + z_{i2}$$
$$\delta_{H_3} = 7.38 + z_{i3}$$
$$\delta_{H_4} = 7.75 + z_{i4}$$
$$\delta_{H_5} = 7.38 + z_{i5}$$
$$\delta_{H_6} = 8.59 + z_{i6}$$

2- or 6-Substituent (i = 2 or 6)	$z_{23} = z_{65}$	$z_{24} = z_{64}$	$z_{25} = z_{63}$	$z_{26} = z_{62}$
-H	0	0	0	0
$-CH_3$	-0.11	-0.01	-0.16	0.08
$-CH_2CH_3$	-0.09	-0.08	-0.15	0.03
$-CH_2$-phenyl	0.12	-0.08	-0.20	0.02
$-CH_2OH$	0.37	0.30	0.02	0.06
$-CH_2NH_2$	0.20	0.07	-0.09	0.05
$-CH_2SC_3H_7$	0.04	-0.08	-0.26	-0.06
$-CH_2SO_2$-Phenyl	$\sim$0	$\sim$-0.3	$\sim$0	-0.2
$-CH=CH_2$	0.11	-0.14	-0.11	0.04
-phenyl	0.16	-0.28	-0.40	-0.03
-2-pyridyl	1.12	-0.09	-0.26	0.00
-F	-0.10	0.40	0.12	-0.13
-Cl	0.32	0.29	0.29	0.20
-Br	0.41	0.17	0.19	0.02
-OH	-0.7	0.0	-1.0	-0.9
$-O-n-C_4H_9$	-0.53	-0.03	-0.49	-0.32
$-NH_2$	-0.68	-0.31	-0.78	-0.48
$-NHCOCH_3$	0.94	0.16	-0.20	-0.10
$-NHCOOCH_2CH_3$	0.59	0.07	-0.24	-0.21
$-NHNO_2$	0.34	0.31	-0.03	-0.41
$-NO_2$	1.09	0.67	0.74	0.26
-CHO	0.93	0.42	0.50	0.44
$-COCH_3$	0.82	0.37	0.39	0.28
-CO-phenyl	0.62	0.55	0.32	0.28
-COOH	0.97	0.43	0.48	0.42
$-COO-n-C_4H_9$	0.86	0.39	0.35	0.35
$-CONH_2$	1.05	0.47	0.43	0.30
$-CSNH_2$	1.41	0.37	0.33	0.25
-CH=NOH	0.40	0.28	0.01	0.16
-CN	0.88	0.38	0.55	0.39

H315

3- or 5-Substituent (i = 3 or 5)	$z_{32} = z_{56}$	$z_{34} = z_{54}$	$z_{35} = z_{53}$	$z_{36} = z_{52}$
-H	0	0	0	0
-CH$_3$	-0.02	-0.06	-0.09	-0.02
-CH$_2$OH	0.11	0.15	0.04	-0.04
-CH$_2$NH$_2$	0.16	0.13	0.04	0.00
-CH$_2$SO$_2$-phenyl	-0.24	-0.15	-0.22	0.01
-CH=CHCOOH	0.45	0.52	0.34	0.17
-Cl	0.20	0.24	0.19	0.09
-Br	0.20	0.43	0.34	0.18
-OH	-0.03	-0.37	0.15	-0.24
-NH$_2$	-0.06	-0.49	0.02	-0.36
-NHCOCH$_3$	0.37	0.50	0.06	-0.16
-SO$_3$H	0.70	1.14	0.81	0.70
-CHO	0.45	0.42	0.12	0.20
-COCH$_3$	0.72	0.68	0.30	0.37
-CO-phenyl	0.47	0.54	0.37	0.34
-COOCH$_3$	0.62	0.60	0.23	0.34
-CSNH$_2$	0.68	0.67	0.24	0.26
-CH=NOH	0.39	0.43	0.19	0.15
-CN	0.63	0.72	0.43	0.50

(row-label groupings at left: C, HAL, N, O=C)

4-Substituent (i = 4)	$z_{42} = z_{46}$	$z_{43} = z_{45}$
-H	0	0
-CH$_3$	0.01	-0.10
-CH$_2$-phenyl	0.00	-0.15
-CH$_2$OH	0.07	0.14
-CH$_2$NH$_2$	0.01	0.03
-CH$_2$S-n-C$_3$H$_7$	-0.06	-0.13
-CH$_2$SO$_2$-phenyl	-0.09	-0.18
-CH=CH$_2$	0.12	0.13
-Cl	0.00	0.05
-Br	0.09	0.35
-OCH$_3$	0.02	-0.29
-NH$_2$	-0.15	-0.74
-NHCOCH$_3$	-0.05	0.31
-SCH$_2$-phenyl	-0.02	0.04
-S-phenyl	0.05	-0.16
-CHO	0.47	0.58
-COCH$_3$	0.40	0.58
-CO-phenyl	0.36	0.40
-COO-n-C$_4$H$_9$	0.34	0.54
-CSNH$_2$	0.35	0.68
-CH=NOH	0.24	0.37
-CN	0.46	0.62

(row-label groupings at left: C, N, S, O=C)

^{1}H-NMR

^{1}H-Chemical Shifts (δ in ppm relative to TMS) and Coupling Constants (J in Hz) in Condensed Heterocycles

(a)	6.66	J_{ab} = 2.5	J_{ce} = 0.9
(b)	7.52	J_{ac} = 0.9	J_{cf} = 0.8
(c)	7.42	J_{ad} = J_{ae} = J_{af} = 0	J_{de} = 7.3
(d)	7.19	J_{bc} = J_{bd} = J_{be} = J_{bf} = 0	J_{df} = 1.2
(e)	7.13	J_{cd} = 8.4	J_{ef} = 7.9
(f)	7.49		

(a)	6.45	J_{ab} = 3.1	J_{de} = 8.1
(b)	7.26	J_{ac} = 2.0	J_{df} = 1.3
(c)	10.1	J_{ad} = 0.7	J_{dg} = 0.9
(d)	7.40	J_{ae} = J_{af} = J_{ag} = 0	J_{ef} = 7.1
(e)	7.09	J_{bc} = 2.5	J_{eg} = 1.2
(f)	6.99	J_{bd} = J_{be} = J_{bf} = J_{bg} = 0	J_{fg} = 7.8
(g)	7.55		

(a)	7.29	J_{ab} = 6	J_{ce} = 2.5
(b)	7.40	J_{ac} = $\sim$0.8	J_{cf} = 0.7
(c)	7.86	J_{ad} = J_{ae} = J_{af} = 0	J_{de} = 7.2
(d)	7.31	J_{bc} = J_{bd} = J_{be} = J_{bf} = 0	J_{df} = 2.5
(e)	7.33	J_{cd} = 9	J_{ef} = 9
(f)	7.78		

(a)	8.08	J_{bc} = J_{de} = 8.2	J_{be} = 0.7
(b,e)	7.70	J_{bd} = J_{ce} = 1.4	J_{cd} = 7.1
(c,d)	7.26		

(a)	8.20	J_{ac} = 0.8	J_{cf} = 1.0
(b)	12.4	J_{ad} = J_{ae} = J_{af} = 0	J_{de} = 7.0
(c)	7.60	J_{cd} = 7.9	J_{df} = 1.2
(d)	7.34	J_{ce} = 1.2	J_{ef} = 7.8
(e)	7.20		
(f)	7.85		

(a)   8.1

(b)
(c) } 7.2–7.9
(d)
(e)

(a)   9.23   in derivatives: $J_{bc} = 8$
(b)   8.23 $J_{bd} = 1.5$
(c)   7.55 $J_{be} = 0.5$
(d)   7.55 $J_{cd} = 7.5$
(e)   8.12 $J_{ce} = 1.5$
                                  $J_{de} = 8$

(a,d) 7.98 $J_{ab} = 8.6$
(b,c) 7.45 $J_{ac} = 0.8$
                $J_{ad} = 1.0$

(a,d) 7.64 $J_{ab} = 9.3$
(b,c) 7.24 $J_{ac} = 1.2$
                $J_{ad} = 0.9$
                $J_{bc} = 6.4$

(a,d) 7.96 $J_{ab} = 9.2$
(b,c) 7.52 $J_{ac} = 1.0$
                $J_{ad} = 0.8$
                $J_{bc} = {\sim}6$

(a)   6.28 $J_{ab} = 3.9$ $J_{de} = 6.8$
(b)   6.64 $J_{ac} = 1.2$ $J_{df} = 1.0$
(c)   7.14 $J_{ad} = 1.0$ $J_{dg} = 1.2$
(d)   7.76 $J_{bc} = 2.7$ $J_{ef} = 6.4$
(e)   6.31 $J_{be} = 0.5$ $J_{eg} = 1.0$
(f)   6.50 $J_{cg} = 0.5$ $J_{fg} = 9.0$
(g)   7.25

^{1}H-NMR

CONDENSED HETEROAROMATICS

(a) 6.38 J_{ab} = 2.2 J_{cf} = 1.0
(b) 7.80 J_{ac} = 0.9 J_{de} = 7.0
(c) 8.39 J_{be} = 0.5 J_{df} = 1.2
(d) 6.62 J_{cd} = 6.9 J_{ef} = 8.9
(e) 6.97 J_{ce} = 1.0
(f) 7.44

(a) 7.48 J_{ab} = 1.0 J_{cf} = 1.0
(b) 7.48 J_{bc} = 0.7 J_{de} = 6.8
(c) 8.09 J_{cd} = 6.9 J_{df} = 1.0
(d) 6.65 J_{ce} = 1.2 J_{ef} = 9.3
(e) 7.03
(f) 7.51

(a) 8.68
(b) 11
(c) 8.99
(d) 9.19

(a) 8.00 J_{ab} = 8.3 J_{df} = 1.1
(b) 7.26 J_{ac} = 1.8 J_{dg} = 0.5
(c) 8.81 J_{ad} = 0.8 J_{ef} = 6.8
(d) 8.05 J_{bc} = 4.3 J_{eg} = 1.6
(e) 7.61 J_{de} = 8.2 J_{fg} = 8.2
(f) 7.43
(g) 7.68

(a) 7.74 J_{ab} = 8.5
(b) 7.27 J_{ac} = 1.1
(c) 8.57 J_{bc} = 6.0
(d) 8.75

H335

(a)	7.50	$J_{ab} = 6.0$		$J_{df} = 1.3$	
(b)	8.45	$J_{ad} = 0.8$		$J_{dg} = 0.9$	
(c)	9.15	$J_{bc} = 0.8$		$J_{ef} = 7.0$	
(d)	7.87	$J_{cg} < 0.5$		$J_{eg} = 1.1$	
(e)	7.50	$J_{de} = 8.2$		$J_{fg} = 8.7$	
(f)	7.57				
(g)	7.71				

(b)	8.14	$J_{ab} = 7.0$	
(c)	8.77	$J_{bc} = 1.7$	

(a)	7.73	$J_{ab} = 5.7$		$J_{cf} = 0.8$	
(b)	9.10	$J_{ac} = 0.8$		$J_{de} = 6.9$	
(c)	8.30	$J_{cd} = 8.6$		$J_{df} = 1.5$	
(d)	7.57	$J_{ce} = 1.3$		$J_{ef} = 7.8$	
(e)	7.57				
(f)	7.57				

(a)	9.29	$J_{ab} = 0$		$J_{cf} = 0.8$	
(b)	9.23	$J_{ac} = 0.5$		$J_{de} = 6.9$	
(c)	8.01	$J_{cd} = 8.5$		$J_{df} = 1.2$	
(d)	7.83	$J_{ce} = 1.2$		$J_{ef} = 7.9$	
(e)	7.58				
(f)	7.84				

(a,b)	8.74	$J_{ab} = 1.8$		$J_{de} = 6.9$	
(c,f)	8.07	$J_{cd} = 8.4$		$J_{df} = 1.6$	
(d,e)	7.68	$J_{ce} = 1.6$		$J_{ef} = 8.4$	
		$J_{cf} = 0.6$			

H340

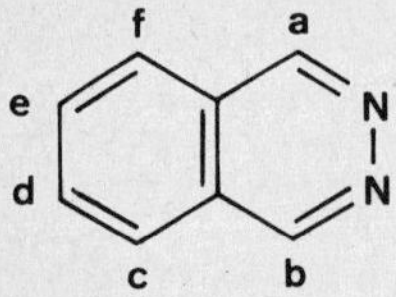

^{1}H-NMR

CONDENSED HETEROAROMATICS

(a,b)	9.44	$J_{ac} = 0.4$		$J_{cf} = 0.6$	
(c,f)	7.93	$J_{cd} = 8.2$		$J_{de} = 6.8$	
(d,e)	7.85	$J_{ce} = 1.2$			

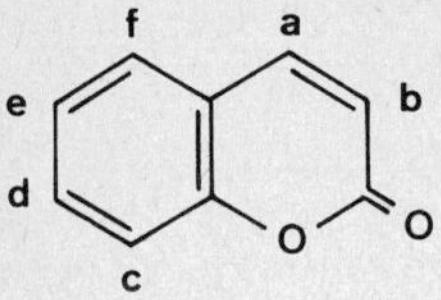

(a)	7.80	$J_{ab} = 9.8$	$J_{de} = 8.6$
(b)	6.45	$J_{cd} = 8.5$	$J_{df} = 2.0$
(c)	7.20	$J_{ce} = 1.8$	$J_{ef} = 8.5$
(d)	7.45	$J_{cf} = 0.0$	
(e)	7.22		
(f)	7.63		

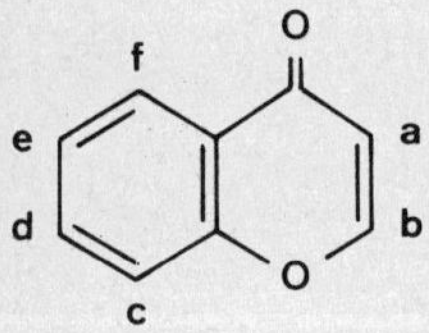

(a)	6.34	$J_{ab} = 6.1$	$J_{de} = 7.0$
(b)	7.88	$J_{cd} = 8.4$	$J_{df} = 1.8$
(c)	7.47	$J_{ce} = 1.1$	$J_{ef} = 8.0$
(d)	7.68	$J_{cf} = 0.5$	
(e)	7.43		
(f)	8.21		

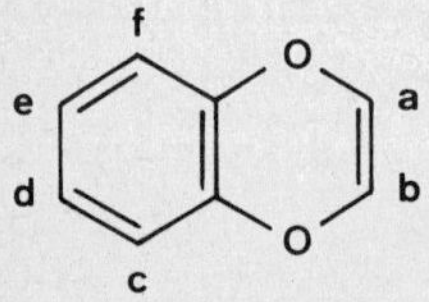

(a,b)	5.77	$J_{cd} = 7.9$
(c,f)	6.52	$J_{ce} = 1.5$
(d,e)	6.71	$J_{cf} = 0.4$
		$J_{de} = 7.9$

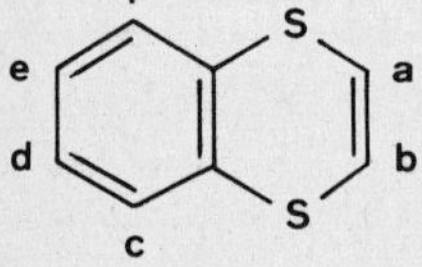

(a,b)	6.42	$J_{cd} = 7.8$
(c,f)	7.19	$J_{ce} = 1.3$
(d,e)	7.12	$J_{cf} = 1.1$
		$J_{de} = 7.1$

(a) 7.84 J_{ab} = 7.6 J_{bc} = 7.3
(b) 7.23 J_{ac} = 1.3 J_{bd} = 0.9
(c) 7.35 J_{ad} = 0.6 J_{cd} = 8.5
(d) 7.48

(a) 8.08 J_{ab} = 7.8 J_{bc} = 7.2
(b) 7.16 J_{ac} = 1.2 J_{bd} = 0.9
(c) 7.36 J_{ad} = 0.7 J_{cd} = 8.2
(d) 7.49 J_{ae} = 0.7
(e) 10.3

(a) 8.36 J_{ab} = 8.0 J_{bc} = 7.1
(b) 7.38 J_{ac} = 1.8 J_{bd} = 1.1
(c) 7.73 J_{ad} = 0.5 J_{cd} = 8.4
(d) 7.50

(a) 8.27 J_{ab} = 8.2 J_{bc} = 7.0
(b) 7.27 J_{ac} = 1.4 J_{bd} = 1.0
(c) 7.74 J_{ad} = 0.4 J_{cd} = 8.6
(d) 7.57 J_{ae} = 0.4
(e) 11.7

(a) 9.09 J_{ab} = 0.4 J_{be} = 0.6
(b) 8.19 J_{ae} = 0.9 J_{cd} = 6.6
(c) 7.64 J_{bc} = 8.2 J_{ce} = 1.2
(d) 7.89 J_{bd} = 1.4 J_{de} = 9.0
(e) 8.22

^{1}H-NMR

Coupling Constants (J in Hz) between ^{1}H and ^{19}F (Spin Quantum Number I = 1/2, Natural Abundance 100 %)

$$\overset{c}{C}H-\overset{b}{C}H-\overset{a}{C}H-F$$

$|J|_{aF}$ = 45–80 (generally 45–50)

$|J|_{bF}$ = 0–30 conformation-dependent:
    synclinal: 0–5
    antiperiplanar: 10–25

$|J|_{cF}$ = 0–4

CH_3F $|J|$ = 81

$\overset{b}{C}H_3\overset{a}{C}H_2F$ $|J|_{aF}$ = 46.7 $|J|_{bF}$ = 25.2

$$\underset{H_c}{\overset{H_b}{>}}C=C\underset{F}{\overset{H_a}{<}}$$

$|J|_{aF}$ = 85 (in derivatives: 70–90)

$|J|_{bF}$ = 52 (in derivatives: 10–50)

$|J|_{cF}$ = 20 (in derivatives: –3...+20)

$H-C\equiv C-F$ $|J|$ = 21

$|J|_{F_{ax}H_{ax}}$ = 34 $|J|_{F_{eq}H_{eq}}$ < 8

$|J|_{F_{ax}H_{eq}}$ = 12 $|J|_{F_{eq}H_{ax}}$ < 8

$|J|_{oF}$ = 9.0 (in derivatives: 6–10)

$|J|_{mF}$ = 5.7 (in derivatives: 4–8)

$|J|_{pF}$ = 0.2 (in derivatives: 0–3)

$|J|_{ortho}$ = 2.5

$|J|_{meta}$ = 0.0

$|J|_{para}$ = 1.5

Coupling Constants ($|J|$ in Hz) Between ^{1}H and ^{31}P (Spin Quantum Number I = 1/2, Natural Abundance 100 %)

$>$P-H $|J|$ = 180-200

$(CH_3CH_2O)_2P\overset{O}{\underset{H}{\big\backslash}}$ $|J|_{aP}$ = 630

$\overset{a\ \ b}{(CH_3CH_2)_3P}$ $|J|_{aP}$ = 13.7
$|J|_{bP}$ = 0.5

$[(CH_3)_2N]_3P$ $|J|$ = 8.8

$\overset{a\ \ b}{(CH_3CH_2)_4P^+}$ $|J|_{aP}$ = 18.0
$|J|_{bP}$ = 13.0

$[(CH_3)_2N]_3PO$ $|J|$ = 9.5

$\overset{a\ \ b}{(CH_3CH_2)_3P=O}$ $|J|_{aP}$ = 16.3
$|J|_{bP}$ = 11.9

$\overset{b}{H}\diagdown\ \ \diagup\overset{a}{H}$ C=C $H\diagup\ \ \diagdown P$, c $|J|_{aP}$ = 10-40
$|J|_{bP}$ = 30-60
$|J|_{cP}$ = 10-30

$\overset{a\ \ b}{(CH_3CH_2O)_3P=O}$ $|J|_{aP}$ = 0.8
$|J|_{bP}$ = 8.4

$\overset{b}{H}\diagdown\ \ \diagup\overset{a}{H}$ C=C $H\diagup\ \ \diagdown O-P$, c $|J|_{aP}$ = ∿7
$|J|_{bP}$ = ∿3
$|J|_{cP}$ = ∿1

$CH_3O\diagdown\overset{O}{\underset{\|}{P}}$, $CH_3O\diagup$ phenyl (o, m, p) $|J|_{oP}$ = 13.3
$|J|_{mP}$ = 4.1
$|J|_{pP}$ = 1.2

¹H-NMR

SOLVENTS

¹H–NMR Spectra of the Most Common Solvents

(δ in ppm relative to TMS)

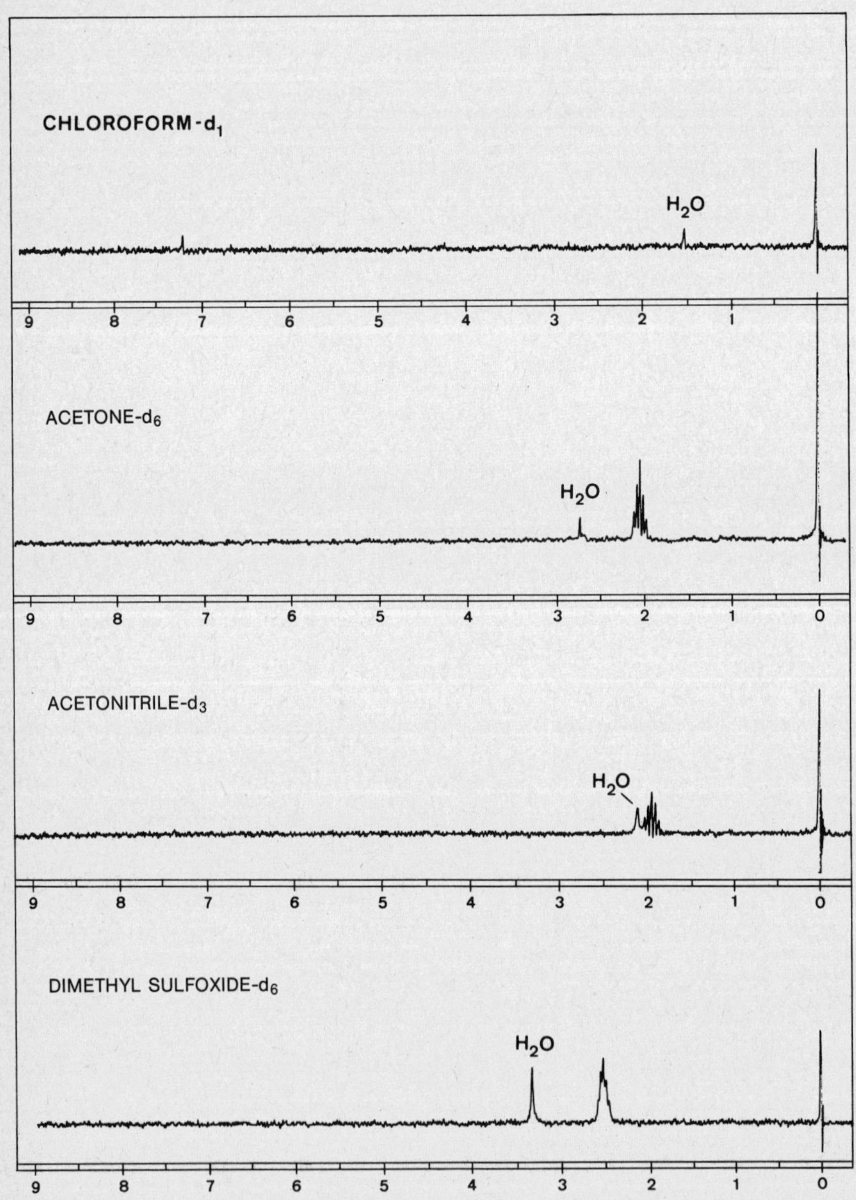

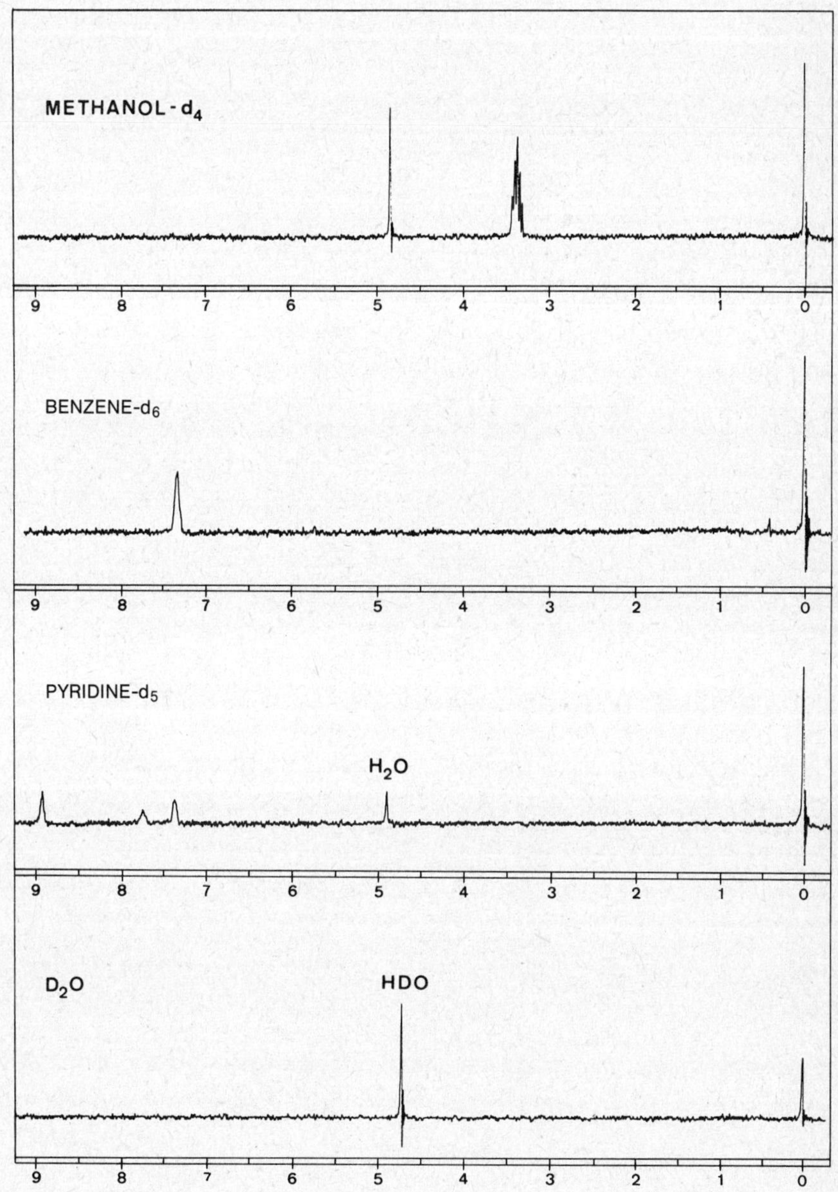
METHANOL-d₄
BENZENE-d₆
PYRIDINE-d₅
H₂O
D₂O
HDO

Characteristic IR-Absorption Bands of Alkyl Groups (in cm^{-1})

Assignment	Range	Comments

$\gtrless$C-H st	3000 - 2840	intensity variable, frequently multiplet
		beyond the normal range:
		    O-CH$_3$ (methyl ethers): 2850-2815
		    O-CH$_2$ (ethers): 2880-2835
		    O-CH$_2$-O (methylenedioxy: 2880-2835, 2780-2750
		    O-CH-O (acetals): ∿2820, weak
		    O◁ , N◁ : 3050-3000
		    CHO (aldehydes): 2900-2800, 2780-2680 (Fermi resonance)
		    for amines: N-CH$_3$, often also for N-CH$_2$-: 2820-2780; in
		        aliphatic dimethylamines in addition 2780-2760
		    △ : 3100-3050, 3035-2995
		    ⬡(cyclohexanes): in addition ∿2700, weak (comb)
		hal-C-H st: 3080-2900

-CH$_3$ δ as	1470 - 1430	medium; coincides with CH$_2$ δ
		beyond the normal range:
		    CH$_3$CO (methyl ketones, acetals): $\Big\}$ 1440-1400
		    CH$_3$-C=C:

$-CH_2$ δ	1475 – 1450	medium, coincides with CH_3 δ as beyond the normal range: $\quad CH_2-C=C$, $CH_2-C\equiv C$: $\sim$1440 $\quad CH_2-C=O$, $CH_2-C\equiv N$, CH_2-X (X: NO_2, halogen, S, P): $\sim$1425
$-CH_3$ δ sy	1395 – 1365	medium doublet in compounds with geminal methyl groups: $\quad -CH(CH_3)_2$: $\sim$1385, $\sim$1370, equal intensity (γ: 1175–1140, doublet) $\quad \geq C(CH_3)_2$: $\sim$1385, $\sim$1365, 1385 weaker than 1365 (γ: 1220–1190, frequently doublet) $\quad -C(CH_3)_3$: $\sim$1390, $\sim$1365, equal intensity, sometimes triplet (γ: 1250–1200, doublet, often incompletely resolved) $\quad -N(CH_3)_2$: no doublet $\quad$ solid-state spectra sometimes exhibit a doublet in the absence of geminal methyl groups beyond the normal range: $\quad O_2S-CH_3$: 1325–1310 $\quad -S-CH_3$ (sulfides): 1330–1290 $\quad P-CH_3$: 1310–1280 $\quad Si-CH_3$: 1275–1260, strong, sharp band

Characteristic IR-Absorption Bands of Alkyl Groups (in cm^{-1})
(continued)

Assignment	Range	Comments

$-CH_3$ γ	1250 - 800	intensity variable; no practical significance
		strong band in compounds with geminal methyl groups:
		   $-CH(CH_3)_2$: 1175-1140, doublet
		   $\geq\!C(CH_3)_2$: 1220-1190, generally doublet
		   $-C(CH_3)_3$: 1250-1200, doublet, often not resolved
		beyond the normal range:
		   $\geq\!SiCH_3$: ∿765; $\geq\!Si(CH_3)_2$: ∿855, ∿800; $-Si(CH_3)_3$: ∿840, ∿765

$-CH_2$ γ	770 - 720	medium, sometimes doublet
		   $C-(CH_2)_n-C$: for n≥4 at ∿720
		             for n<4 at higher wave numbers, weaker
		                in cyclohexanes at ∿890
		beyond the normal range:
		   cycloalkanes: 1060-800, a number of bands; unreliable

Characteristic IR-Absorption Bands of Alkenyl Groups (in cm^{-1})

Assignment	Range	Comments
$=CH_2$ st	3095 – 3075	medium, often a number of bands
$\diagup\!\!\!=$CH st	3040 – 3010	CH st in aromatics and three-membered rings fall in the same range; in cyclic compounds: ∿3075 ∿3060 ∿3045 ∿3020
=CH δ ip	1420 – 1290	no practical significance

Characteristic IR-Absorption Bands of Alkenyl Groups (in cm^{-1})

(continued)

Assignment	Range	Comments			
=CH δ oop	1005 - 675	a number of bands sub-ranges:			
			C=C	C=C-C=O	C=C-OR	C=C-O-C=O

	C=C	C=C-C=O	C=C-OR	C=C-O-C=O
$-CH=CH_2$:	1005–985	∿980	∿960	∿950
	920–900	∿960	∿815	∿870
	(also overtone at 1850–1800)	∿810		
$>C=CH_2$:	900–880	∿940	∿795	
	(also overtone at 1850–1780)	∿810		
C=C (cis):	990–960	∿975	∿960	∿950
C=C (trans):	725–675	∿820		
C=C:	840–800	∿820		

in the same range also arC-H δ oop, C-O-C γ and C-N-C γ in saturated heterocycles, OH δ oop in carboxylic acids, NH γ, N-O st, S-O st, CH_2 γ, C-F δ (?), C-Cl st

C=C st	1690 - 1635	variable intensity; weak if compound highly symmetric, strong for

N-C=C and O-C=C

sub-ranges:

$-CH=CH_2$: 1650-1635

$\diagdown C=CH_2$: 1660-1640

H$\diagdown$C=C$\diagup$H : 1690-1665, weak

H$\diagdown$C=C$\diagup$H : 1665-1635

$\diagdown$C=C$\diagup$H : 1690-1660 ⎫

$\diagdown$C=C$\diagdown$: 1690-1650 ⎭ weak, often absent

beyond the normal range:

C=C-X, X = O, N, S: at lower frequency (to ~1590),
increased intensity; in vinyl ethers often doublet
due to rotational isomers

at lower frequency if conjugated with:

C=C	C≡C	C≡N	(benzene)	C=O	(cyclopropyl)	(epoxide)
~1650 ~1600	~1600	~1620	~1630	~1630	~1640	~1640

Characteristic IR-Absorption Bands of Alkenyl Groups (in cm^{-1})

(continued)

Assignment	Range	Comments
		in and on small rings (C=C st):  ~1640 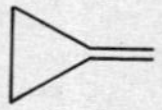 ~1780 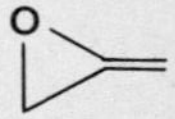 ~1650  ~1570  ~1640  ~1680 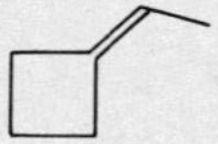 ~1690  ~1610 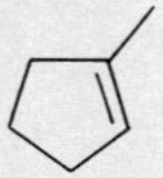 ~1660 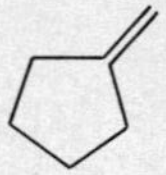 ~1660 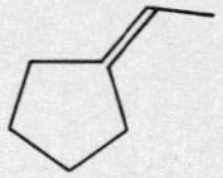 ~1670 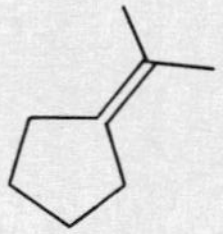 ~1690  ~1570  ~1650 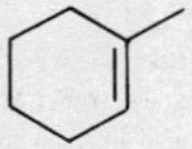 ~1675 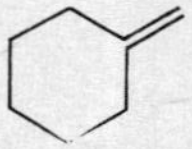 ~] 50 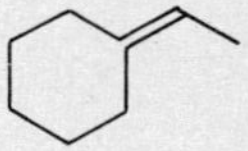 ~1665 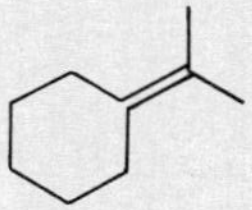 ~1670  ~1615

Characteristic IR-Absorption Bands of Alkynyl Groups (in cm^{-1})

Assignment	Range	Comments
$\equiv$C–H st	3340 – 3250	strong, sharp in the same range also OH st and NH st
C$\equiv$C st	2260 – 2100	weak, sharp R–C$\equiv$C–H: at the lower end of the range cited R–C$\equiv$C–R: usually 2 bands (Fermi resonance); often missing if substitution symmetrical sub-ranges: C–C$\equiv$C–H C–C$\equiv$C–C C–C$\equiv$C–C$\equiv$N C–C$\equiv$C–COOH C–C$\equiv$C–COOCH$_3$ ~2120 ~2220 ~2240 ~2240 ~2240,2140 in the same range also C$\equiv$Z st, X=Y=Z st, Si–H st
$\equiv$C–H δ	700 – 600	weak, no practical significance

Characteristic IR-Absorption Bands of Aromatic Compounds (in cm^{-1})

Assignment	Range	Comments
arC-H st	3080 – 3030	often a number of bands of lower intensity in the same range also CH st of alkenes and small rings
arC-C	1625 – 1575	medium, often doublet; generally weak in benzene derivatives with a center of symmetry in the ring in the same range also C=C st, C=N st, C=O st, N=O st, C-C in heterocycles, NH δ
	1525 – 1475	medium, often doublet; weak for in the same range also N=O st, C-C in heterocycles, C=O st, B-N st, CH_3 δ, CH_2 δ, NH δ
comb	2000 – 1650	very weak; useful for the determination of substitution pattern in 6-membered aromatic rings, see p. I55 in the same range also H_2O δ, C=O st, B-H$\cdots$B st, $\overset{+}{N}$-H st
arC-H δ ip	1250 – 950	a number of bands of variable intensity, no practical significance

| arC-H δ oop | 900 - 650 | one or more strong bands; in 6-membered aromatic rings useful for the determination of substitution pattern, see p. I55 in the same range also =C-H δ oop, C-O-C γ and C-N-C γ in saturated heterocycles, OH δ oop in carboxylic acids, NH δ, N-O st, S-O st, CH$_2$ γ, C-F δ (?), C-Cl st |

5-ring-heteroaromatics:

	pyrrole (N-H)	furan (O)	thiophene (S)
NH st free:	3500-3400	—	—
hydrogen bonded:	3400-2800		
CH st	∿3100	∿3100	∿3100
γ	1590-1560	1610-1560	1535-1515
intensity variable, generally multiplets	1540-1500	1510-1475	1455-1410
CH δ oop generally strong	770- 710	990- 725	935- 700

Characteristic IR-Absorption Bands of Aromatic Compounds (in cm^{-1})

(continued)

Determination of substitution pattern in 6-membered aromatic rings:

not to be used for ring systems with strongly conjugated substituents such as $C=O$, NO_2, $C{\equiv}N$

position and shape of bands related to the number of adjacent H-atoms

number of adjacent H	type of substitution in benzene derivatives	band positions	
		comb	δ oop, γ
5	mono-	2000 1600	~900, 770–730, 710–690
4	o-di-		770–735
(1 + 3)	m-di-		900–860, 865–810*, 810–750, 725–680

		2000 1600	
3	vic-tri-		800–770, 720–685, 780–760*
(1 + 2)	asym-tri-		900–860, 860–800, 730–690
2	p-di- 1,2,3,4-tetra-		860–780
1	1,3,5-tri- 1,2,3,5-tetra- 1,2,4,5-tetra- penta-		900–840; in 1,3,5-trisubstituted benzenes also: 850–800, 730–675*
0	hexa-		

*: band sometimes missing

Characteristic IR-Absorption Bands of Compounds of the Type $X \equiv Y$ (in cm^{-1})

(Nitriles, Cyanates, Thiocyanates, Isonitriles, Diazo Compounds)

Assignment	Range	Comments
$-C \equiv N$ st	2260 – 2240	intensity medium to strong, sharp. For $O-CH_2-C \equiv N$ and $N-CH_2-C \equiv N$ lower intensity beyond the normal range: $C=C-C \equiv N$, ⬡$-C \equiv N$: 2240–2215 $XC-C \equiv N$, X : Cl, Br, I: 2240–2230; $-CF_2-C \equiv N$: ∿2275 $>N-C \equiv N \longleftrightarrow >\overset{+}{N}=C=\overset{-}{N}$: 2225–2175 $>N-C=C-C \equiv N$: 2210–2185 $C \equiv N \rightarrow O$: 2305–2280; $N=O$ st: 1395–1365 $(C \equiv N)^-$: 2200–2070 $(OC \equiv N)^-$: 2220–2130; $(OC \equiv N)^-$ st sy: 1335–1290 $-S-C \equiv N$: 2170–2135; C–S st: 725–550, often doublet $(SC \equiv N)^-$: 2090–2020; C–S st: 750
$-\overset{+}{N} \equiv \overset{-}{C}$	2150 – 2110	strong
$-\overset{+}{N} \equiv N$	2310 – 2130	medium; frequency depends on anion in the region for $X \equiv Y$ st also $C \equiv C$ st, $X=Y=Z$ st as, $\overset{+}{N}H$ st, PH st, POH st, SiH st, BH st

Characteristic IR-Absorption Bands of Compounds of the Type X=Y=Z (in cm^{-1})

(Allenes, Ketenes, Ketenimines, Diazo Compounds, Isocyanates, Isothiocyanates, Carbodiimides, Azides)

Assignment	Range	Comments
(C=C)=C-H st	3050 - 2950	in this range also CH st of other compounds
C=C=C st as	1950 - 1930	strong, doublet in $X-C=C=CH_2$ if X not alkyl
		ring strain increases frequency: $\triangleright=CH_2$: ~2020
		for $C=C=CH_2$: overtone at ~1700, weak
		C=C=C st sy: 1075-1060, weak, absent if high symmetry
(C=C)=CH_2 δ oop	~850	strong
C=C=O st as	2155 - 2130	very strong, CH_2=C=O: ~2155; $(\text{C}_6\text{H}_5)_2$-C=C=O: ~2130
C=C=N st as	2050 - 2000	very strong, sometimes doublet

Characteristic IR-Absorption Bands of Compounds of the Type X=Y=Z (in cm^{-1})

(Allenes, Ketenes, Ketenimines, Diazo Compounds, Isocyanates, Isothiocyanates, Carbodiimides, Azides)

(continued)

Assignment	Range	Comments

$\overset{+}{C}=\overset{-}{N}=N$ st as	2050 – 2010	very strong
		sub-ranges:
		$R-CH=\overset{+}{N}=\overset{-}{N}$: 2050–2035
		$R_2-C=\overset{+}{N}=\overset{-}{N}$: 2035–2010 $\biggr\}$ R = ar or al
		beyond the normal range:
		$R-CO-C=\overset{+}{N}=\overset{-}{N}$: 2100–2050
		    C=O st: R = al: ∿1645; R = ar: ∿1615
		    $C=\overset{+}{N}=\overset{-}{N}$ st sy: ∿1350, strong
		$O=\!\!\bigcirc\!\!=\overset{+}{N}=\overset{-}{N}$: 2180–2010
		    C=O st: 1645–1560

–N=C=O st as	2275 – 2230	strong, sharp; CH$_3$NCO: ∿2230; $\bigcirc$–NCO: ∿2275; –CF$_2$NCO: ∿2300
		    N=C=O st sy: ∿1390, weak
		beyond the normal range:
		    (N=C=O)$^{-}$: 2220–2130

-N=C=S st as	2150 - 2050	very strong, generally doublet (Fermi resonance) R-N=C=S st sy: R = al: $\sim$950; R = ar: 700-650 beyond the normal range: (N=C=S)$^-$: 2090-2020
-N=C=N- st as	2150 - 2100	very strong, doublet if aromatic substituent -N=C=N st sy: $\sim$1390, very weak
$\overset{+}{-}\overset{-}{N}=N=N$ st as	2250 - 2080	very strong, in acyl azides often doublet (C=O st: $\sim$1700) $-\overset{+}{N}=N=\overset{-}{N}$ st sy: 1350-1180, strong other related systems: C=S=O st: $\sim$1100, $\sim$1050 N=S=O st: $\sim$1250, $\sim$1135 in the region for X=Y=Z st as also C$\equiv$C st, X$\equiv$Y st, $\overset{+}{N}$H st, PH st, POH st, SiH st, BH st

Characteristic IR-Absorption Bands of Alcohols and Phenols (in cm^{-1})

Assignment	Range	Comments

-OH st	3650 - 3200	intensity variable
		sub-ranges:
		   -OH free: 3650-3590, sharp
		   -OH hydrogen bonded: 3550-3450, broad
		   -OH polymer: 3500-3200, broad, often a number of bands
		
		beyond the normal range:
		   R-O-H $\longleftrightarrow$ R=O$\cdots$H (chelates): 3200-2500, often very broad
		in the same range also NH st, $\equiv$CH st: $\sim$3300, sharp, H_2O

-OH δ ip	1450 - 1200	medium, no practical significance

C-O st	1260 - 970	strong, often doublet
		sub-ranges:
		   $-CH_2-OH$: 1075-1000
		   $\rangle$CH-OH : 1125-1000
		   $\geq$C-OH : 1210-1100
		   arC-OH : 1260-1180
		in this range also bands for C-F st, C-N st, N-O st, P-O st,
		    C=S st, S=O st, P=O st, Si-O st, Si-H δ

-OH δ oop	<700	medium, no practical significance

Characteristic IR-Absorption Bands of Ethers (in cm^{-1})

Assignment	Range	Comments
C-O-C st as	1310 - 1000	strong, sometimes split (C-O-C st sy weak)

sub-ranges for non-cyclic ethers:

$-CH_2-O-CH_2-$: 1150-1085 (OCH_3 st: 2850-2815;

OCH_2 st: 2880-2835)

$>CH-O-CH<$: 1170-1115, often split

C=C-O-C-al : 1225-1180 (st sy: 1125-1080, medium)

arC-O-C-al : 1275-1200 (st sy: 1075-1020, medium)

sub-ranges for cyclic ethers

: ~1270, ~840 (CH st: 3050-3000)

: ~1030, ~980

: ~1070, ~915; : ~1235

: ~1100, ~815

ketals, acetals: 4 to 5 bands in the range indicated,

in cyclic compounds also:

: ~950; : ~925 (CH_2 st: ~2780)

: ~880; : ~800

in acetals CH st: ~2820, weak

strong bonds in the same range also for C-O st, C-F st,

C-N st, N-O st, P-O st, C=S st, S=O st, P=O st, Si-O

st, Si-H δ

Characteristic IR-Absorption Bands of Peroxides and Hydroperoxides (in cm^{-1})

Assignment	Range	Comments

O-O-H st	3450 - 3200	intensity variable
		sub-ranges:
		  -OOH free: ∿3450
		  -OOH hydrogen bonded: about 30 cm^{-1} higher than in the
		    corresponding alcohols; in peracids at ∿3300
		in the same range also OH st, NH st, ≡CH st, H_2O

C-O-O st	1200 - 1000	strong, about 20 cm^{-1} lower than in the corresponding alcohols
		strong bands in the same range also for C-O st, C-F st,
		  C-N st, N-O st, P-O st, C=S st, S=O st, P=O st, Si-O st,
		  Si-H δ

O-O st	1000 - 800	medium or weak; assignment uncertain
		
		also:
		  C=O st in peracids: 1760-1745
		    in diacylperoxides: 1820-1770, two bands

Characteristic IR-Absorption Bands of Amines (in cm^{-1})

Assignment	Range	Comments
$-NH_2$ st	3500 - 3300	intensity variable, generally 2 sharp bands at lower wave numbers (often <3200) if hydrogen bonded, broader hydrogen bonded and free form often simultaneously observed in primary aromatic amines an additional band at $\sim$3200 in the same range also OH st, $\equiv$CH st
$\rangle$NH st	3450 - 3300	intensity variable, only one band at lower wave numbers (often <3200) if hydrogen bonded, broader hydrogen bonded and free form often simultaneously observed in the same range also OH st, $\equiv$CH st: $\sim$3300, sharp, H_2O
$-\overset{+}{N}H_3$ st $\rangle\overset{+}{N}H_2$ st $\geq\overset{+}{N}H$ st	3000 - 2000	medium, broad, very structured band sub-ranges for major maximum: $\overset{+}{N}H_3$: 3000-2700; also $\overset{+}{N}H_3$ comb: $\sim$2000 $\overset{+}{N}H_2$: 3000-2700 $\overset{+}{N}H$: 2700-2250 in the same range also OH st, NH st, CH st, SH st, PH st, SiH st, BH st, X=Y=Z st, X$\equiv$Y st

Characteristic IR-Absorption Bands of Amines (in cm^{-1})

(continued)

Assignment	Range	Comments
$-NH_2\ \delta$ $>NH\ \delta$	1650 – 1550	medium or weak
$\overset{+}{-}NH_3\ \delta$ $\overset{+}{>}NH_2\ \delta$ $\overset{+}{\geqq}NH\ \delta$	1600 – 1460	medium, weak in aliphatic amines, often more than one band
C–N st	1400 – 1000	medium, no practical significance
$-NH_2\ \delta$ $>NH\ \delta$	850 – 700	medium or weak, in primary amines 2 bands
		also: P–N–C st: 1110–930, 770–680

Characteristic IR-Absorption Bands of Halogen Compounds (in cm^{-1})

Assignment	Range	Comments
C-F st	1400 - 1000	strong, sharp, often more than one band (rotational isomers), often not resolved sub-ranges: alCF : 1100-1000 } FC-H st: 3080-2990 alCF$_2$: 1150-1000 alCF$_3$: 1350-1100 C=CF : 1350-1150; C=CF$_2$ st: ∿1745 arCF : 1250-1100 strong bands in the same range also for C-O st, NO$_2$ st sy, C=S st, S=O st
CF$_2$ CF$_3$	780 - 680	medium or weak, assignment uncertain (C-F δ?) also: S-F st : 815-755, strong P-F st : 1110-760 Si-F st: 980-820 B-F st : 1500-800

<u>Characteristic IR-Absorption Bands of Halogen Compounds</u> (in cm^{-1})

(continued)

Assignment	Range	Comments
C-Cl st	830 - <600	strong, often broad (rotational isomers); absent in aryl chlorides; in -CCl$_3$: 830-700
C-Br st	<700	
C-I st	<600	

in disubstituted halobenzenes characteristic skeletal vibration:

type of substitution	o	m	p
X : Cl	1055-1035	1080-1075	1095-1090
X : Br	1045-1030	1075-1065	1075-1070
X : I			1060-1055

also:

P-Cl st : <600

Si-Cl st: <625

B-Cl st : 1100-650

Characteristic IR-Absorption Bands of Aldehydes (in cm^{-1})

Assignment	Range	Comments
comb	2900 - 2800 2780 - 2680	weak (Fermi resonance with C-H δ at ∿1390. For extreme position of C-H δ only one band) sub-ranges: al: 2830-2810; 2720-2690 ar: 2830-2810; 2750-2720, for o-substitution often higher in the same range also a weak band of cyclohexanes: ∿2700
C=O st	1765 - 1645	strong sub-ranges: al: 1740-1720; α-halogenated: 1765-1730 ar: 1710-1685; αβ-unsaturated: 1695-1660 with intramolecular H-bonds: 1670-1645

Characteristic IR-Absorption Bands of Ketones (in cm^{-1})

Assignment	Range	Comments

C=O st	1775 - 1650	strong
		sub-ranges:
		  al: ∿1715; branching at α-position shifts to lower
		    frequencies: >C< (C=O) : ∿1695; +C+ (C=O) : ∿1685
		cyclic ketones:
		  ∿1775 ∿1750 ∿1715 $(CH_2)_{6-8}$ C=O ∿1705 $(CH_2)_{14}$ C=O ∿1715
		conjugated ketones:
		  αβ-unsaturated: ∿1675, often 2 bands (rotational isomers);
		    C=C st: 1650-1600 (▷-C-al (C=O) : ∿1695)
		  αβ,γδ-unsaturated:
		  αβ,α'β'-unsaturated: ∿1665 (▷-C-C=C (C=O): ∿1670)
		aryl ketones: ∿1690 (▷-C-ar (C=O): ∿1675)
		diaryl ketones: ∿1665 (with N or O in p-position to ∿1600)

α-halogenated ketones: shifted towards higher wave numbers,
depending on dihedral angle ϕ between C=O and C-hal;
largest effect for $\phi = 0°$, no effect for $\phi = 90°$

maximal shift:

α-chloro: ∿25

α,α-dichloro: ⎫
α,α'-dichloro: ⎬ ∿45

α-bromo: ∿20

α-iodo: ∿0

α,α-difluoro: ∿60

perfluoro: ∿90

diketones:

α-diketones: al: ∿1720; 5-ring: ∿1775, ∿1760; 6-ring: ∿1760,
∿1730; enolized: ∿1675 (C=C st: ∿1650)
ar: ∿1680; ortho-quinones: ∿1675, with peri-OH:
∿1675, ∿1630

β-diketones: keto form: ∿1720 (sometimes doublet); enol form:
∿1650, with intramolecular H-bonds:
∿1615 (C=C st: ∿1600, strong)

γ-diketones: like normal ketones; para-quinones: ∿1675 (C=C st:
∿1600, with peri-OH: ∿1675, ∿1630

Characteristic IR-Absorption Bands of Esters and Lactones (in cm^{-1})

Assignment	Range	Comments

C=O st	1790 – 1650	strong
		sub-ranges:
		    aliphatic esters: 1750-1735
		    conjugated esters:
		        αβ-unsaturated acid: 1730-1710
		        ar acid: 1730-1715
		        with intramolecular H-bonds: 1690-1670
		        α-halogenated acid: 1790-1740
		        vinyl esters: ∿1760 (C=C st: 1690-1650, strong)
		        phenol esters: ∿1760
		        phenol esters of an ar acid: ∿1735
		diesters: like the corresponding monoesters
		keto esters:
		    α-keto: 1755-1725, generally only one band
		    β-keto: keto form: ∿1750 (ketone), ∿1735 (ester)
		            enol form: ∿1650 (C=C st: ∿1630, strong)
		    γ-keto: ∿1740, ∿1715; pseudo esters: ∿1770

lactones:

~1840

~1770   ~1800   ~1750

(additional band at ~1780
if α-position free)

~1735   ~1760   ~1720   ~1730

(often doublet)

C-O st	1330 - 1050	2 bands: st as: very strong and at higher wave number; st sy: strong, at lower wave number

sub-ranges for st as:

formates, propionates, higher al esters: ~1185

acetates: ~1240

vinyl esters, phenol esters: ~1210

γ-lactones, δ-lactones: ~1180

methyl esters of al acids: ~1165

in the same range also strong bands for C-F st, C-N st,
N-O st, P-O st, C=S st, S=O st, P=O st, Si-O st,
Si-H δ

IR

Characteristic IR-Absorption Bands of Amides, Lactams, Imides and Hydrazides (in cm^{-1})

Assignment	Range	Comments

N-H st	3500 - 3100	medium, in primary amides 2 bands, in polypeptides and proteins multiplet
		sub-ranges:
		  free: 3500-3400
		  hydrogen bonded: 3350-3100
		    in primary amides generally 2 bands at ∿3350,3180
		    in lactams generally 2 bands at ∿3200,3100
		    in hydrazides: mono-: ∿3200; di-: ∿3100
		    in imides: ∿3250
		in the same range also OH st, ≡CH st: ∿3300, sharp, H_2O

C=O st (amide I)	1740 - 1630	generally strong

sub-range:

amides:	H$_2$NCO-	-HNCO-	>NCO-
free:	∿1690	∿1685	∿1650
hydrogen bonded:	∿1650	∿1660	∿1650

lactams: 4-ring: ∿1745
         5-ring: ∿1700
         6,7-ring: ∿1650

hydrazides: mono-: ∿1670; di-: ∿1600

imides: 1740-1670; in 5-ring imides 2 bands at ∿1750,1700

polypeptides: 1655-1630

isocyanurates: ∿1690, at ∿1770 for aromatic substitution

trifluoroacetates: ∿1720, 1755 (shoulder)

NH δ / N-C=O st sy (amide II)	1630 - 1510	generally strong

sub-range:

	H$_2$NCO-	-HNCO-	lactams
free:	∿1610	∿1530	} absent
hydrogen bonded:	∿1630	∿1540	

polypeptides: 1560-1510

trifluoroacetates: ∿1555

also:	H$_2$NCO-	-HNCO-	lactams
C-N st (?)	∿1400	∿1250	∿1330
NH δ ip	∿1150		∿1465
NH δ oop	750-600	∿700	∿800

IR

$$\overset{X}{\underset{Y}{\diagdown}}C=Z$$

Characteristic <u>IR-Absorption Bands of Carbonic Acid Derivatives</u> (in cm^{-1})

Assignment	Range	Comments
C=O st	1820 – 1630	strong sub-ranges: al–O–CO–O–al: ∿1740 (C–O st as: ∿1260) ar–O–CO–O–al: ∿1770 (C–O st as: ∿1230) ar–O–CO–O–ar: ∿1785 (C–O st as: ∿1215) $\begin{bmatrix}O\\O\end{bmatrix}$C=O : ∿1820 (C–O st as: ∿1200) al–S–CO–O–al: ∿1705 (C–O st as: ∿1150) ar–S–CO–O–al: ∿1725 (C–O st as: ∿1130) al–S–CO–O–ar: ∿1735 (C–O st as: ∿1075) al–S–CO–S–al: ∿1645 (C–S st: ∿880) ar–S–CO–S–al: ∿1650 (C–S st: ∿840) ar–S–CO–S–ar: ∿1715 (C–S st: ∿830) al–O–CO–Cl : ∿1780 (C–O st: ∿1155) ar–O–CO–Cl : ∿1785 al–S–CO–Cl : ∿1770 ar–S–CO–Cl : ∿1775

$$\underset{Y}{\overset{X}{{\diagdown}}}C{=}Z$$

		$H_2N{-}CO{-}O{-}$: ~1700 (NH_2 δ: ~1620)
		$-NH{-}CO{-}O{-}$: ~1725 (C–N st: ~1230)
		$H_2N{-}CO{-}S{-}$: ~1700 (NH_2 : ~1620)
		$-NH{-}CO{-}S{-}$: ~1670 (C–N st: ~1190)
		$\rangle N{-}CO{-}S{-}$: ~1670 (C–N st: ~1250)
		$\rangle N{-}CO{-}Cl$: ~1740
		$-NH{-}CO{-}NH{-}$: ~1660 (NH >O: 1780–1680, 3 bands)
C=N st	1725 – 1625	strong
		sub-ranges:
		$-O{-}C({=}NH){-}O{-}$: 1690–1645 (C–O st: ~1325,1100; NH st: 3400–3200)
		+ +
		$-O{-}C({=}NH_2){-}O{-}$: 1680–1635 (NH_2 δ: 1590–1540)
		$\rangle N{-}C({=}NH){-}N\langle$: 1725–1625 (NH st: 3400–3200; NH δ: 1670–1620)
		$-NH{-}C({=}NR){-}O{-}$: 1675–1655
C=S st	1225 – 1030	strong
		sub-ranges:
		$O{-}CS{-}O$: ~1225
		$O{-}CS{-}N$: ~1170
		$N{-}CS{-}N$: 1340–1130
		$S{-}CS{-}N$: ~1050
		$S{-}CS{-}S$: ~1070

Characteristic IR-Absorption Bands of Carboxylic Acids (in cm^{-1})

Assignment	Range	Comments

COO-H st	3550 - 2500	intensity variable
		sub-ranges:
		    COO-H st free: 3550-3500, sharp; only in highly dilute solutions
		    COO-H st hydrogen bonded: 3300-2500, broad, often more than one band
		in the same range also OH st, NH st, CH st, SiH st, SH st, PH st

C=O st	1800 - 1650	strong
		    free: 1800-1740
		    hydrogen bonded (dimer): 1740-1650 } also in dicarboxylic acids
		sub-ranges for hydrogen bonded C=O:
		    al-COOH: 1725-1700
		    C=C-COOH: 1715-1690
		    ar-COOH: 1700-1680
		    hal-C-COOH: 1740-1720
		    intramolecular H-bonded: 1670-1650

OC–OH st ⎫ OC–OH δ ⎬	1440 – 1210	no practical significance
OC–OH δ oop	960 – 880	medium, generally broad (only in dimers) in the same range also =CH δ, arCH δ, NH δ
$(COO)^-$ st as	1610 – 1550	very strong; in α-halogen carboxylates near the higher value, with more than one α-halogen beyond the normal range. in polypeptides at ∿1575
$(COO)^-$ st sy	1450 – 1400	strong, no practical significance; in polypeptides at ∿1470

also:

$(COO)^-$ δ: formates: ∿775, weak

            acetates: ∿925

           benzoates: ∿680

        CF_3COO^-:   ∿600

Characteristic IR-Absorption Bands of Amino Acids (in cm^{-1})

Assignment	Range	Comments

N-H st
O-H st
	3400 - 2000	generally strong, broad, very structured
		sub-ranges:
		    zwitter ion: 3100-2000, distinct side band at 2200-2000
		    hydrochloride: 3350-2000
		    Na-salt: 3400-3200

$\overset{+}{N}H_3\ \delta$ as	1660 - 1590	weak; for hydrochloride near the lower value
$\overset{+}{N}H_3\ \delta$ sy	1550 - 1480	medium

COO^- st as	1760 - 1595	strong
		sub-ranges:
		    zwitter ion: ∿1595
		    hydrochloride: 1755-1700, in α-amino acids: 1760-1730
		    Na-salt: ∿1595

Characteristic IR-Absorption Bands of Acid Halides (in cm^{-1})

Assignment	Range	Comments
C=O st	1820 – 1750	strong; the range is for chlorides; fluorides higher (1900–1870), bromides and iodides lower in aromatic acid chlorides and bromides an additional band at ∿1725 (Fermi resonance)
$-C{\overset{=O}{\underset{Hal}{}}}$ (?)	1300 – 900	strong; assignment uncertain in aromatic acid chlorides and bromides an additional band at ∿870

IR

Characteristic IR-Absorption Bands of Acid Anhydrides (in cm^{-1})

Assignment	Range	Comments

C=O st	1870 – 1725	strong, 2 bands in this range
		sub-ranges:
		    linear anhydrides: ∿1820, ∿1760, higher band stronger
		    cyclic anhydrides:
		        5-ring: ∿1850, ∿1775 }
		        6-ring: ∿1800, ∿1760 } lower band stronger

C-O-C st as	1050 – 900	strong; linear: ∿1040, cyclic: ∿920

Characteristic IR-Absorption Bands of Compounds with C=N Groups (in cm^{-1})

(Azomethines, Schiff's Bases, Imines, Amidines, Azines, Imino Carbonates)

Assignment	Range	Comments
C=N st	1690 - 1580	generally strong: sub-ranges: R-CH=N-R': R, R' = al: $\sim$1670; R or R' = conjugated: $\sim$1645; R, R' = conjugated: $\sim$1630 $CH_3COCH=N$-⬡ : C=O st: $\sim$1655; C=N st: $\sim$1555 $\frac{R}{R''}$C=N-R': R, R', R" = al: $\sim$1655; R = conjugated: $\sim$1645; R, R' = conjugated: $\sim$1635 $\frac{R}{R'}$C=NH : R, R' = al: $\sim$1645; R = conjugated: $\sim$1625 R-C=N-R : 1685-1580; in R-C=N-R additional 1540-1515 	| NH_2 NHR CH=N-N=CH: 1670-1600 $\frac{RO}{RO}$C=NH : 1690-1645 (NH st: $\sim$3300; C-O st: $\sim$1325, $\sim$1100) $\frac{RO}{RO}$C=$\overset{+}{N}H_2$: 1680-1635 ($\overset{+}{N}H_2$ st: $\sim$3000; $\overset{+}{N}H_2$ δ: 1590-1540) S-C=N : 1640-1605; S-S-C=N: 1640-1580 O-C=N : 1690-1645 also: P=N st: 1500-1170

Characteristic IR-Absorption Bands of Oximes (in cm^{-1})

Assignment	Range	Comments

OH st	3600 - 2700	strong
		sub-ranges:
		    free: ∿3600
		    hydrogen-bonded: 3300-3100, broad
		    quinone oximes: to ∿2700, more than one band

C=N st	1685 - 1520	generally strong
		sub-ranges:
		    al: 1685-1650
		    ar: 1645-1650
		    quinone oximes: 1580-1520 (C=O st: 1680-1620)

OH δ	1475 - 1315	no practical significance
N-O st	1050 - 400	

Characteristic IR-Absorption Bands of Compounds with N=N Groups (in cm^{-1})

(Azo Compounds, Azoxy Compounds, Azothio Compounds)

Assignment	Range	Comments
N=N st	1500 - 1400	very weak, missing in compounds of high symmetry
$-N=N \nearrow^{O}_{\searrow}$ st as	1480 - 1450	
st sy	1335 - 1315	
$-N=N \nearrow^{S}_{\searrow}$ st as	~1450	
st sy	~1050	
$O \searrow^{}_{}N=N \nearrow^{O}_{\searrow} O$ st	1410 - 1175	sub-ranges:
		trans cis
		al: 1290-1175 1425-1385, 1345-1320
		ar: 1300-1250 ~1410, ~1395
		(dimers of C-nitroso compounds; monomers:
		al: 1585-1540; ar: 1510-1490)

Characteristic IR-Absorption Bands of Nitrites and Nitroso Compounds (in cm^{-1})

Assignment	Range	Comments

N=O st	1680 – 1450	very strong
		sub-ranges:
		   O-NO (nitrites): trans: 1680–1650; cis: 1625–1610
		   C-NO (C-nitroso compounds): al: 1585–1540; ar: 1510–1490
		       (dimers: see p. I200)
		   N-NO (N-nitroso compounds): ∿1450
		
		in nitrites also: comb: 3300–3200, ∿2500, 2300–2250
		              N-O st: trans: ∿800; cis: very weak
		            ONO δ: trans: ∿600; cis: ∿650
		in C-nitroso compounds also: C-N st: al: ∿850; ar: ∿1100
		in N-nitroso compounds also: N-N st: ∿1040

Characteristic IR-Absorption Bands of Nitrates, Nitro Compounds and Nitramines (in cm^{-1})

Assignment	Range	Comments
NO_2 st as	1660 – 1490	very strong sub-ranges: $O-NO_2$ (nitrates): 1660–1625 $C-NO_2$ (nitro compounds): al: 1570–1540; ar: 1560–1490 $N-NO_2$ (nitramines): 1630–1530
NO_2 st sy	1390 – 1260	strong sub-ranges: $O-NO_2$ (nitrates): 1285–1270 $C-NO_2$ (nitro compounds): al: 1390–1340 ar: 1360–1310, often 2 bands $N-NO_2$ (nitramines): 1315–1260 in nitrates also: N-O st: $\sim$870, strong; NO_2 γ: $\sim$760, NO_2 δ: $\sim$700

Characteristic IR-Absorption Bands of Mercaptans, Thioethers and Disulfides (in cm^{-1})

Assignment	Range	Comments
S-H st	2600 - 2540	often weak, narrow
S-H δ	915 - 800	weak, no practical significance
C-S st	710 - 570	weak, broad, no practical significance
S-S st	∿500	weak, no practical significance
		also:
		$(S-)CH_3$ st as: ∿2880; $(S-)CH_2$ st as: ∿2860
		$(S-)CH_3$ δ as: ∿1430, δ sy: 1330-1290; $(S-)CH_2$ δ: ∿1425
		S-F st: 815-755, strong
		S-N st in S-N=O: ∿630
		S-C st in S-C≡N: 725-550, often doublet

Characteristic IR-Absorption Bands of Compounds with C=S Groups (in cm^{-1})

(Thioketones, Thioesters, Dithioacids, Thioamides, Thiolactams)

Assignment	Range	Comments

C=S st	1275 - 1030	strong, narrow
		sub-ranges:
		   thioketones: 1075-1030
		   thioesters: 1210-1080
		   dithioacids: ∿1215 (SH st: ∿2550; SH δ: ∿860)
		   thioacid fluoride: 1125-1075 (perfluorinated: 1130-1105)
		   thioacid chloride: 1100-1065 (perfluorinated: 1100-1075)
		   thioamides:
		   thiolactams: } 1140-1090 (C-N st: 1535-1520; NH δ: 1380-1300)
		
		also:
		   P=S st: 750-580

Characteristic IR-Absorption Bands of Compounds with SO Groups (in cm^{-1})

Assignment	Range	Comments

$\rangle$S=O st	1225 - 980	strong
		sub-ranges:
		   R-SO-R : 1060-1015
		   R-SO-OH : ∿1100 (S-O st: 870-810; OH st: free: ∿3700;
		       hydrogen bonded: ∿2900, ∿2500)
		   R-SO-OR : ∿1135 (S-O st: 740-720, 710-690)
		   RO-SO-OR: 1225-1195
		   R-SO-Cl : ∿1135
		   $R-SO_2^-$: ∿1030, ∿980
		   R=SO : ∿1100, ∿1050 (N=SO: ∿1250, ∿1135)

| $\overset{\diagdown}{\underset{\diagup}{S}}\!\!\overset{\textstyle =\!O}{\underset{\textstyle =\!O}{}}$ st as / st sy | 1420 – 1000 | very strong
sub-ranges:
   $R-SO_2-R$: 1370–1290, 1170–1110
   $R-SO_2-OR$: 1375–1350, 1185–1165
   $R-SO_2-SR$: $\sim$1340, $\sim$1150
   $RO-SO_2-OR$: 1415–1390, 1200–1185
   $R-SO_2-N$: 1365–1315, 1180–1150 (NH st: 3330–3250;
      NH δ: $\sim$1750; SN st: 725–650)
   $R-SO_2-hal$: 1410–1375, 1205–1170
   $R-SO_2-OH$: 1355–1340, 1165–1150 (OH st: $\sim$2900, $\sim$2400;
      hydrated: 2800–1650, broad, S=O st like for $R-SO_3^-$)
   $R-SO_3^-$: 1250–1140, 1070–1030
   $RO-SO_3^-$: 1315–1220, 1140–1050 |
| S–O st | 870 – 690 | variable intensity, in sulfites weak |

Characteristic IR-Absorption Bands of Phosphorous Compounds (in cm^{-1})

Assignment	Range	Comments
P–H st	2440 – 2275	weak to medium, generally only 1 band, in $R_3\overset{+}{P}H$ very broad
PO–H st	2700 – 2650	weak, very broad bands
POH comb	2300 – 2250	in $P{<}^{O}_{OH}$ also 1740–1600 (dimer?)
P–O st	1260 – 855	sub-ranges: P–O–C st: al: 1050–970, strong; 830–740, often weak ar: 1260–1160; $P^{\underline{V}}$: 995–915; $P^{\underline{III}}$: 875–855 P–OH st : 1100–940, broad; for $P(OH)_2$ often 2 bands P–O–P st: 980–900
P=O st	1300 – 960	strong sub-ranges: $R_3P{=}O$: 1190–1150 $R_2(R'O)P{=}O$: 1265–1200 } also for R = H $R(R'O)_2P{=}O$: 1280–1240 $(RO)_3P{=}O$: 1300–1260 $R(HO)_2P{=}O$: 1220–1150; $R(HO)PO_2^-$: 1250–990, more than one band; RPO_3^{2-} : 1125–970, 1000–960

$R_2(HO)P{=}O$: 1205–1090; $R_2PO_2^-$: 1200–1090, 1090–995

$RO(HO)_2P{=}O$: ~1250; $RO(HO)PO_2^-$: 1230–1210, 1030–1020;

$ROPO_3^{2-}$: 1140–1050, 1010–970

$(RO)_2(HO)P{=}O$: 1250–1210; $(RO)_2PO_2^-$: 1285–1120, 1120–1050

$R(RO)(HO)P{=}O$: 1220–1170; $R(RO)PO_2^-$: 1245–1150, 1110–1050

$\begin{matrix} R_2P{=}O \\ \quad\diagdown O \\ R_2P{=}O \end{matrix}$: 1240–1205

$\begin{matrix} R(HO)P{=}O \\ \quad\diagdown O \\ R(HO)P{=}O \end{matrix}$: ~1195

$\begin{matrix} R(RO)P{=}O \\ \quad\diagdown O \\ R(RO)P{=}O \end{matrix}$: 1265–1250

$\begin{matrix} RO(HO)P{=}O \\ \quad\diagdown O \\ RO(HO)P{=}O \end{matrix}$: ~1250

$\begin{matrix} (RO)_2P{=}O \\ \quad\diagdown O \\ (RO)_2P{=}O \end{matrix}$: 1310–1260

$\begin{matrix} RO(R_2N)P{=}O \\ \quad\diagdown O \\ RO(R_2N)P{=}O \end{matrix}$: ~1275

$\begin{matrix} (RO)_2P{=}O \\ \quad\diagdown O \\ (R_2N)_2P{=}O \end{matrix}$: ~1300, ~1240

$\begin{matrix} (R_2N)_2P{=}O \\ \quad\diagdown O \\ (R_2N)_2P{=}O \end{matrix}$: ~1235

$R_2(X)P{=}O$: 1265–1240

$R(X)_2P{=}O$: 1365–1260

$(RO)_2(X)P{=}O$: 1330–1280

$RO(X)_2P{=}O$: 1365–1260

for X = F, Cl, Br
(for X = F near the upper
limit of the range)

Characteristic IR-Absorption Bands of Phosphorous Compounds (in cm^{-1})

(continued)

Assignment	Range	Comments
P-OH δ	$\sim$1280	weak, no practical significance
P-C st	800 - 700	intensity varies widely, no practical significance
P-H δ	1090 - 910	strong, for $(RO)_2HP=O$ very strong
		also:
		P-N-C st: 1110-930, 770-680
		P=N-R st: R = al : 1500-1230
		R = ar : 1390-1300
		R = C=O: 1370-1310
		R = PR_2: 1295-1170
		P=S st: 750-580, intensity varies widely (P-S st: <600)
		(P-)CH_3 δ sy: 1310-1280
		P-F st: 905-760 (PF_2: 1110-800, more than one band)
		P-Cl st: <600

Characteristic IR-Absorption Bands of Silicon Compounds (in cm^{-1})

Assignment	Range	Comments
Si-H st	2250 - 2090	medium sub-ranges: R_3SiH: 2160-2090; also for R = H, for SiH_3 2 bands hal-SiH: to ∿2250 (Si-O)SiH: 2220-2120
Si-H δ	1010 - 700	strong, broad, generally 2 bands
(Si-)CH_3 δ sy	1275 - 1260	very strong, sharp, typical for $SiCH_3$; not split for $Si(CH_3)_2$ ($SiCH_3$ δ sy: ∿1410, weak)
(Si-)CH_3 γ	860 - 760	$SiCH_3$: ∿765; $Si(CH_3)_2$: ∿855, ∿800; $Si(CH_3)_3$: ∿840, ∿765
Si-O st	1110 - 1000 900 - <600	Si-O-C : 1110-1000, 850-800 Si-O-Si: 1090-1030, <650 Si-OH : 900- 800 (SiO-H st: 3700-3200; Si-OH δ: ∿1030)
Si-C st	850 - 650	

Characteristic IR-Absorption Bands of Silicon Compounds (in cm^{-1})

(continued)

Assignment	Range	Comments

Si-N st	1250 - 830	sub-ranges:
		    Si-N-Si: 950-830 (Si_2NH st: $\sim$3400)
		    N-Si-N: 950-830
		    $Si-NH_2$: 1250-1100 ($SiN-H_2$ st: $\sim$3570, $\sim$3390; $Si-NH_2$ δ: $\sim$1540)

Si-F st	980 - 820	sub-ranges:
		    Si-F : 920-820
		    $Si-F_2$: 945-870, 2 bands
		    $Si-F_3$: 980-860, 2 bands
		
		also:
		    Si-Cl st: <625

Characteristic IR-Absorption Bands of Boron Compounds (in cm^{-1})

Assignment	Range	Comments
B–H st	2640 – 2200	strong; in B–H···B: 2200–1540, more than one band
B–O st	1380 – 1310	very strong, in haloboroxines to ∿1500 (BO–H st: 3300–3200, very broad)
B–N st	1550 – 1330	very strong

B–C st	1240 – 620	strong, 2 bands if substitution highly asymmetric
		
		also:
		    B–F st : 1500–800
		    B–Cl st: 1100–650

INTERFERENCES, OPAQUE REGIONS,
SUSPENSION MEDIA

Interferences in Infrared Spectra

Traces of water in carbon tetrachloride or chloroform may give rise to two bands in the vicinity of 3700 and 3600 cm^{-1} as well as one around 1600 cm^{-1}. Water in the vapor phase exhibits many sharp bands between 2000 and 1280 cm^{-1}. If present in high concentration these bands may temporarily block the detector of a double beam instrument. This effect can simulate a shoulder if it occurs while scanning through a steep side of a strong signal. Certain shoulders on carbonyl bands can often be rationalized in this way.

In solution, carbon dioxide shows an absorption band at 2325 cm^{-1}. In solutions which contain amines, CO_2 can form carbonates with traces of water, which leads to the appearance of unexpected bands of protonated nitrogen-containing groups. In improperly balanced double beam instruments gaseous CO_2 can give rise to two signals at approximately 2360 and 2335 cm^{-1} as well as a signal at 667 cm^{-1}.

Commercially available polymers often contain phthalates as plasticizers which can be found in apparently pure samples and give rise to a band at 1725 cm^{-1}. In the course of chemical reactions these phthalates can be transformed to phthalic anhydride which gives a band at 1755 cm^{-1}.

Other frequently encountered contaminants are silicones, which generally exhibit a band at 1265 cm^{-1} together with a broad signal in the region from 1100 to 1000 cm^{-1}.

Opaque Regions and Suspension Media

In the figures shown on pages H270, H275 and H280, the regions shown in black along the wave number scale have very low transparency for that particular solvent which may lead to certain artifacts. These regions should be disregarded in the interpretation of the spectra.

The bands of the suspension matrices are always found superimposed on those of the sample.

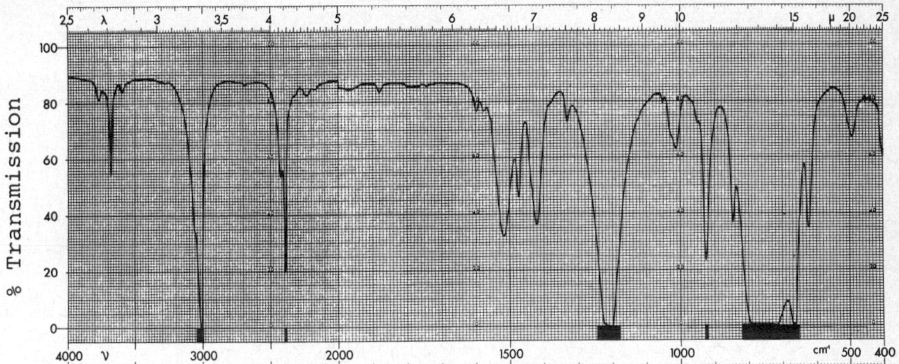

Chloroform: 0.2 mm thickness

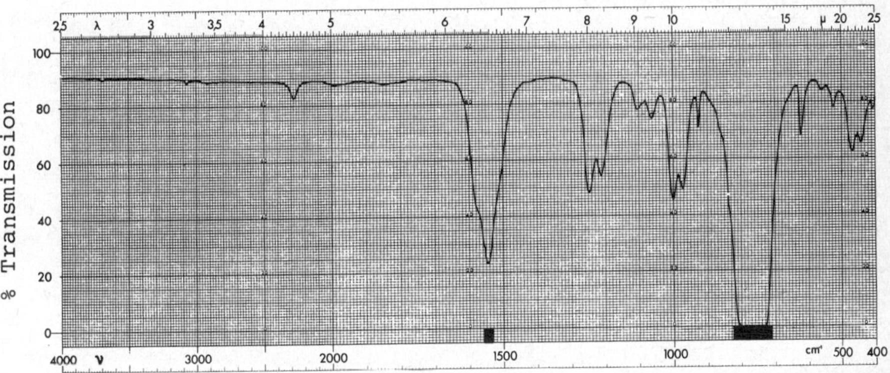

Carbon tetrachloride: 0.2 mm thickness

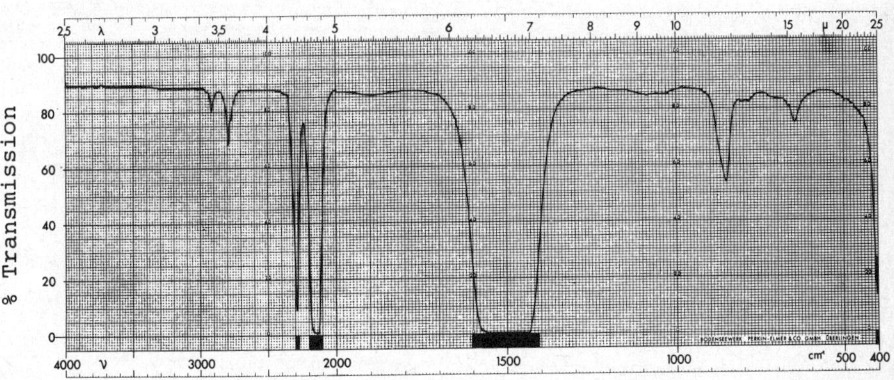

Carbon disulfide: 0.2 mm thickness

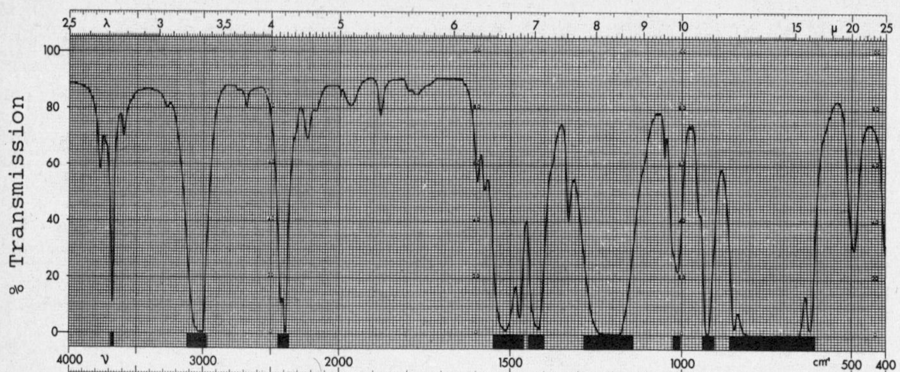

Chloroform: 1.0 mm thickness

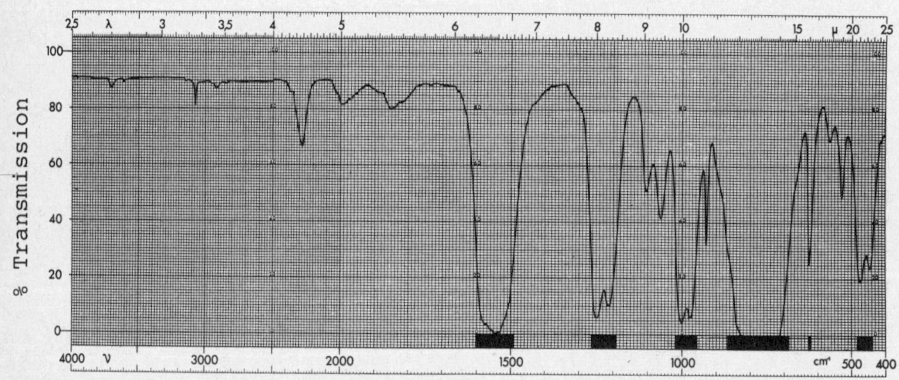

Carbon tetrachloride: 1.0 mm thickness

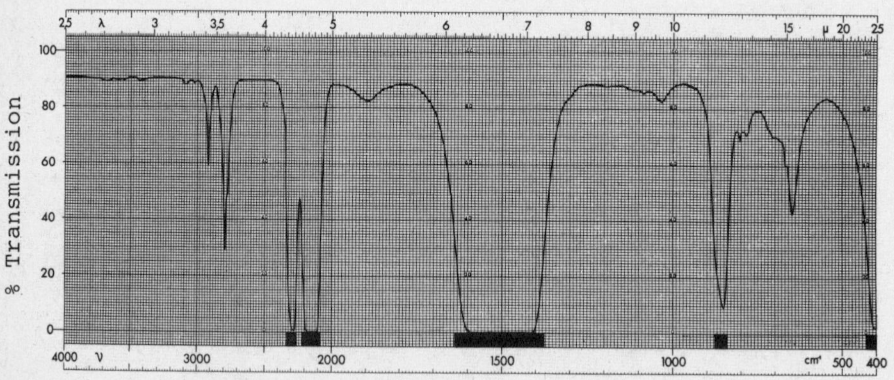

Carbon disulfide: 1.0 mm thickness

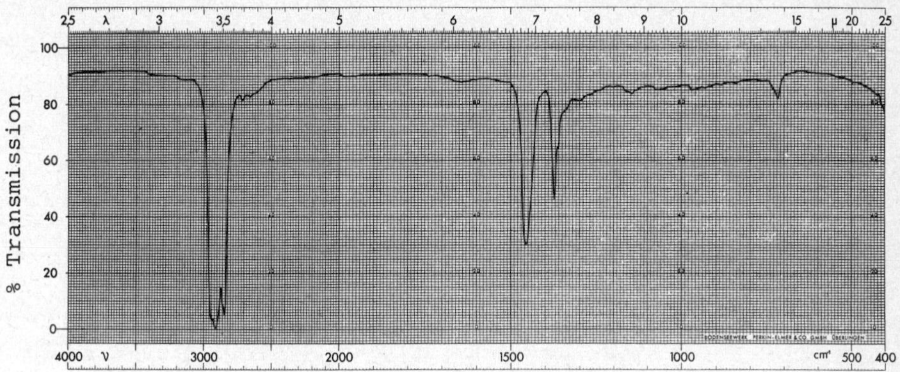

Nujol: ca. 12 μ thickness

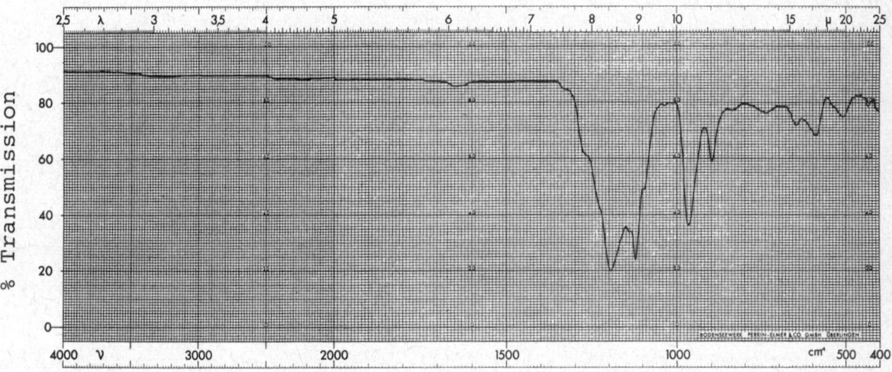

Fluorolube: ca. 12 μ thickness

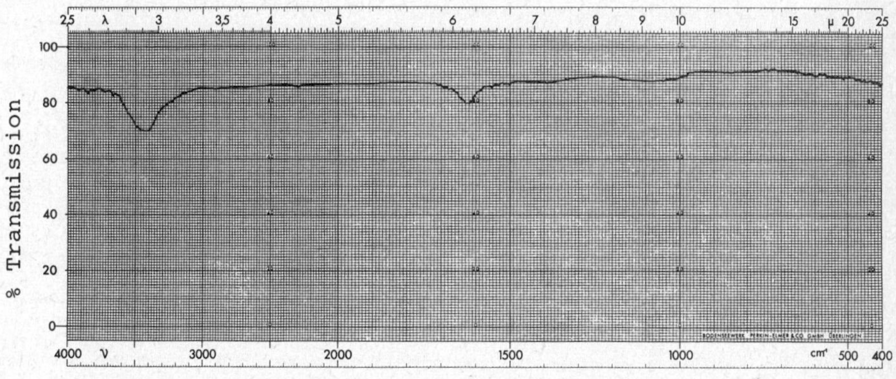

Potassium bromide: pellet

MASS CORRELATION TABLE

<u>Mass Correlation Table</u>

Mass	Ion	Product ion and composition of the neutral particle lost. $M\overset{+}{\cdot}$ = molecular ion	Sub-structure or compound type
12	$C\overset{+}{\cdot}$		
13	CH^+		
14	$CH_2\overset{+}{\cdot}$, N^+, N_2^{++}, CO^{++}		
15	CH_3^+	$M\overset{+}{\cdot}-15$ (CH_3)	non-specific; abundant: methyl, N-ethylamines
16	$O\overset{+}{\cdot}$, NH_2^+, O_2^{++}	$M\overset{+}{\cdot}-16$ (CH_4)	methyl (rare)
		(O)	nitro compounds, sulfones, epoxides, N-oxides
		(NH_2)	primary amines
17	OH^+, $NH_3\overset{+}{\cdot}$	$M\overset{+}{\cdot}-17$ (OH)	acids (especially aromatic acids), hydroxylamines, N-oxides, nitro compounds, sulfoxides, tertiary alcohols
		(NH_3)	primary amines
18	$H_2O\overset{+}{\cdot}$, NH_4^+	$M\overset{+}{\cdot}-18$ (H_2O)	non-specific <u>O-indicator</u> abundant: alcohols, some acids, aldehydes, ketones, lactones, cyclic ethers
19	H_3O^+, F^+	$M\overset{+}{\cdot}-19$ (F)	fluorides <u>F-indicator</u>
20	$HF\overset{+}{\cdot}$, Ar^{++}, CH_2CN^{++}	$M\overset{+}{\cdot}-20$ (HF)	fluorides

21	$C_2H_2O^{++}$		
22	CO_2^{++}		
23	$Na^{+\cdot}$		
24	$C_2^{+\cdot}$		
25	C_2H^+	$M^{+\cdot}-25$ (C_2H)	terminal acetylenyl
26	$C_2H_2^{+\cdot}$, CN^+	$M^{+\cdot}-26$ (C_2H_2)	aromatics
		(CN)	nitriles
27	$C_2H_3^+$, $HCN^{+\cdot}$	$M^{+\cdot}-27$ (C_2H_3)	terminal vinyl, some ethyl esters and N-ethylamides, ethyl phos-phates
		(HCN)	aromatic N, nitriles
28	$C_2H_4^{+\cdot}$, $CO^{+\cdot}$, $N_2^{+\cdot}$, $HCNH^+$	$M^{+\cdot}-28$ (C_2H_4)	non-specific; abundant: cyclo-hex-enes, ethyl esters, propyl ketones, propyl-substituted aromatics
		(CO)	aromatic O, quinones, lactones, lactams, unsaturated cyclic ketones, allyl aldehydes
		(N_2)	diazo compounds
29	$C_2H_5^+$, CHO^+	$M^{+\cdot}-29$ (C_2H_5)	non-specific; abundant: ethyl
		(CHO)	phenols, furans, aldehydes
30	$CH_2O^{+\cdot}$, $CH_2NH_2^+$, NO^+, $C_2H_6^{+\cdot}$ $BF^{+\cdot}$, $N_2H_2^{+\cdot}$ <u>N-indicator</u>	$M^{+\cdot}-30$ (C_2H_6)	ethylalkanes, polymethyl compounds
		(CH_2O)	cyclic ethers, lactones, primary alcohols
		(NO)	nitro and nitroso compounds
31	CH_3O^+, $CH_3NH_2^{+\cdot}$, CF^+, $N_2H_3^+$ <u>O-indicator</u>	$M^{+\cdot}-31$ (CH_3O)	methyl esters, methyl ethers, primary alcohols
		(CH_3NH_2)	N-methylamines

Mass	Ion	Product ion and composition of the neutral particle lost. $M^{+\cdot}$ = molecular ion	Sub-structure or compound type
32	$O_2^{+\cdot}$, $CH_3OH^{+\cdot}$, $S^{+\cdot}$, $N_2H_4^{+\cdot}$ O-indicator		(N_2H_3)	hydrazides
		$M^{+\cdot}-32$ (CH_3OH)	methyl esters, methyl ethers
		(S)	sulfides
		(O_2)	cyclic peroxides
33	$CH_3OH_2^{+}$, SH^{+}, CH_2F^{+}	$M^{+\cdot}-33$ (CH_3 + H_2O)	non-specific O-indicator
		(SH)	non-specific S-indicator
		(CH_2F)	fluoromethyl
34	$SH_2^{+\cdot}$ S-indicator	$M^{+\cdot}-34$ (SH_2)	non-specific S-indicator
		(OH + OH)	nitro compounds
35	SH_3^{+}, Cl^{+}	$M^{+\cdot}-35$ (Cl)	chlorides
		(OH + H_2O)	nitro compounds 2 x O-indicator
36	$HCl^{+\cdot}$, C_3^{+}	$M^{+\cdot}-36$ (HCl)	chlorides
		(H_2O + H_2O)	2 x O-indicator
37	C_3H^{+}		
38	$C_3H_2^{+\cdot}$		
39	$C_3H_3^{+}$	$M^{+\cdot}-39$ (C_3H_3)	aromatics
40	$C_3H_4^{+\cdot}$, CH_2CN^{+}, $Ar^{+\cdot}$	$M^{+\cdot}-40$ (CH_2CN)	cyanomethyl
41	$C_3H_5^{+}$, $CH_3CN^{+\cdot}$	$M^{+\cdot}-41$ (C_3H_5)	alicyclics (especially poly-alicyclics), alkenes
		(CH_3CN)	2-methyl-N-aromatics, N-methyl-anilines

42	$C_3H_6^{+\cdot}$, $C_2H_2O^{+\cdot}$, CON^+, $C_2H_4N^+$	$M^{+\cdot}-42$ (C_3H_6)	non-specific; abundant: propyl esters, butyl ketones, butyl aromatics, methyl cyclohexenes
		(C_2H_2O)	acetates (especially enolacetates), acetamides, cyclohexenones, $\alpha\beta$-unsaturated ketones
43	$C_3H_7^+$, $C_2H_3O^+$, $CONH^{+\cdot}$	$M^{+\cdot}-43$ (C_3H_7)	non-specific; abundant: propyl, cycloalkanes, cycloalkanones, cycloalkylamines, cycloalkanols, butyl aromatics
		(CH_3CO)	methyl ketones, acetates, aromatic methyl ethers
44	$CO_2^{+\cdot}$, $C_2H_6N^+$, $C_2H_4O^+$, $C_3H_8^{+\cdot}$ $CH_4Si^{+\cdot}$	$M^{+\cdot}-44$ (C_3H_8)	propyl alkanes
		(C_2H_6N)	N,N-dimethylamines, N-ethylamines
		(C_2H_4O)	cycloalkanols, cyclic ethers, ethylene ketals
		(CO_2)	anhydrides, lactones, carboxylic acids
45	$C_2H_5O^+$, CHS^+, $C_2H_7N^{+\cdot}$ <u>O-indicator</u>, <u>S-indicator</u>	$M^{+\cdot}-45$ (C_2H_5O)	ethyl esters, ethyl ethers, lactones ethyl sulfonates, ethyl sulfones
		(CHO_2)	carboxylic acids
		(C_2H_7N)	N,N-dimethylamines, N-ethylamines
46	$C_2H_5OH^{+\cdot}$, NO_2^+	$M^{+\cdot}-46$ (C_2H_6O)	ethyl esters, ethyl ethers, ethyl sulfonates
		$(H_2O + C_2H_4)$	primary alcohols
		$(H_2O + CO)$	carboxylic acids
		(NO_2)	nitro compounds
47	CH_3S^+, CCl^+, $C_2H_5OH_2^+$, $CH(OH)_2^+$ <u>S-indicator</u>, <u>2 x O-indicator</u>	$M^{+\cdot}-47$ (CH_3S)	methyl sulfides
48	$CH_3SH^{+\cdot}$, $CHCl^{+\cdot}$, $SO^{+\cdot}$	$M^{+\cdot}-48$ (CH_4S)	methyl sulfides
		(SO)	sulfoxides, sulfones, sulfonates

Mass	Ion	Product ion and composition of the neutral particle lost. $M^{+\cdot}$ = molecular ion	Sub-structure or compound type
49	CH_2Cl^+, $CH_3SH_2^{+\cdot}$	$M^{+\cdot}-49$ (CH_2Cl)	chloromethyl
50	$C_4H_2^{+\cdot}$, $CH_3Cl^{+\cdot}$, $CF_2^{+\cdot}$	$M^{+\cdot}-50$ (CF_2)	trifluoromethyl aromatics, perfluoro alicyclics
51	$C_4H_3^+$, CHF_2^+		
52	$C_4H_4^{+\cdot}$		
53	$C_4H_5^+$		
54	$C_4H_6^{+\cdot}$, $C_2H_4CN^+$	$M^{+\cdot}-54$ (C_4H_6)	cyclohexene
		(C_2H_4CN)	cyanoethyl
55	$C_4H_7^+$, $C_3H_3O^+$	$M^{+\cdot}-55$ (C_4H_7)	non-specific; abundant: cycloalkanes, butyl esters, N-butylamides
56	$C_4H_8^{+\cdot}$, $C_3H_4O^{+\cdot}$	$M^{+\cdot}-56$ (C_4H_8)	butyl esters, N-butylamides, pentyl ketones, cyclohexenes, tetralins, pentylaromatics
		(C_3H_4O)	methylcyclohexenones, β-tetralones
57	$C_4H_9^+$, $C_3H_5O^+$, $C_3H_2F^+$	$M^{+\cdot}-57$ (C_4H_9)	non-specific
		(C_3H_5O)	ethyl ketones
58	$C_3H_6O^{+\cdot}$, $C_3H_8N^+$	$M^{+\cdot}-58$ (C_4H_{10})	alkanes
	<u>N-indicator</u>, <u>O-indicator</u>	(C_3H_6O)	α-methyl alkanals, methyl ketones, isopropylidene glycols
59	$C_3H_7O^+$, $C_2H_5NO^{+\cdot}$	$M^{+\cdot}-59$ (C_3H_7O)	propyl esters, propyl ethers
	<u>O-indicator</u>		$(C_2H_3O_2)$	methyl esters

60	$C_2H_4O_2^{+\bullet}$, $CH_2NO_2^+$, $C_2H_6NO^+$ <u>O-indicator</u>	$M^{+\bullet}-60$	(C_3H_9N)	amines, amides
			(C_3H_8O)	propyl esters, propyl ethers
			$(C_2H_4O_2)$	acetates
			(CH_3OH+CO)	methyl esters
61	$C_2H_5O_2^+$, $C_2H_5S^+$ <u>S-indicator, 2 x O-indicator</u>	$M^{+\bullet}-61$	$(C_2H_5O_2)$	glycols, ethylene ketals
			(C_2H_5S)	ethyl sulfides
62	$C_2H_6O_2^{+\bullet}$, $C_2H_3Cl^{+\bullet}$	$M^{+\bullet}-62$	$(C_2H_6O_2)$	methoxymethyl ethers, ethylene glycols, ethylene ketals
			(C_2H_6S)	ethyl sulfides
63	$C_5H_3^+$, $C_2H_4Cl^+$, $COCl^+$	$M^{+\bullet}-63$	(C_2H_4Cl)	chloroethyl
			$(Cl+CO)$	acid chlorides
64	$C_5H_4^{+\bullet}$, $SO_2^{+\bullet}$, $S_2^{+\bullet}$	$M^{+\bullet}-64$	(SO_2)	sulfones, sulfonates
			(S_2)	disulfides
65	$C_5H_5^+$	$M^{+\bullet}-65$	(S_2H)	disulfides
66	$C_5H_6^{+\bullet}$	$M^{+\bullet}-66$	(C_5H_6)	cyclopentenes
67	$C_5H_7^+$, $C_4H_3O^+$	$M^{+\bullet}-67$	(C_4H_3O)	furyl ketones
68	$C_5H_8^{+\bullet}$, $C_4H_4O^{+\bullet}$, $C_3H_6CN^+$	$M^{+\bullet}-68$	(C_5H_8)	cyclohexenes, tetralins
			(C_4H_4O)	cyclohexenones, β-tetralones
69	$C_5H_9^+$, $C_4H_5O^+$, $C_3HO_2^+$, CF_3^+	$M^{+\bullet}-69$	(C_5H_9)	alicyclics, alkenes
			(CF_3)	trifluoromethyl
70	$C_5H_{10}^{+\bullet}$			alkanes, alkenes, cycloalkanes
	$C_4H_6O^{+\bullet}$			cycloalkanones
	$C_4H_8N^+$			pyrrolidines
71	$C_5H_{11}^+$			alkanes, larger alkyl groups
	$C_4H_7O^+$			alkanones, alkanals, tetrahydrofurans

MASS CORRELATION TABLE

Mass	Ion	Compound type
72	$C_4H_8O^{+\cdot}$	alkanones, alkanals <u>O-indicator</u>
	$C_4H_{10}N^+$	aliphatic amines <u>N-indicator</u>
	C_6^+	perhalogenated benzenes
73	$C_4H_9O^+$	alcohols, ethers, esters <u>O-indicator</u>
	$C_3H_5O_2^+$	acids, esters, lactones
	$C_3H_9Si^+$	trimethylsilyl compounds
74	$C_4H_{10}O^{+\cdot}$	ethers
	$C_3H_6O_2^{+\cdot}$	methyl esters of carboxylic acids, α-methyl carboxylic acids
75	$C_3H_7O_2^+$	methyl acetals, glycols <u>2 x O-indicator</u>
	$C_3H_7S^+$	sulfides, thiols <u>S-indicator</u>
	$C_2H_7SiO^+$	trimethylsiloxyl compounds
76	$C_6H_4^{+\cdot}$	aromatics
77	$C_6H_5^+$	aromatics
	$C_3H_6Cl^+$	chlorides
78	$C_6H_6^{+\cdot}$	aromatics
	$C_5H_4N^+$	pyridines
	$C_3H_7Cl^{+\cdot}$	chlorides
79	$C_6H_7^+$	aromatics with H-containing substituents
	$C_5H_5N^{+\cdot}$	pyridines, pyrroles
	Br^+	bromides
80	$C_6H_8^{+\cdot}$	cyclohexenes, polycyclic alicycles
	$C_5H_4O^{+\cdot}$	cyclopentenones
	$HBr^{+\cdot}$	bromides
	$C_5H_6N^+$	pyrroles, pyridines
81	$C_6H_9^+$	cyclohexanes, cyclohexenyls, dienes
	$C_5H_5O^+$	furans, pyrans
82	$C_6H_{10}^{+\cdot}$	cyclohexanes
	$C_5H_6O^{+\cdot}$	cyclopentenones, dihydropyrans
	$C_5H_8N^+$	tetrahydropyridines
	$C_4H_6N_2^{+\cdot}$	pyrazoles, imidazoles

M40

Mass	Ion	Compound type
83	$C_6H_{11}^+$	alkenes, cycloalkanes, monosubstituted alkanes
	$C_5H_7O^+$	cycloalkanones
84	$C_5H_{10}N^+$	piperidines, N-methylpyrrolidines
85	$C_6H_{13}^+$	alkanes
	$C_5H_9O^+$	alkanones, alkanals, tetrahydropyrans, fatty acid derivatives
86	$C_5H_{10}O^{+\cdot}$	alkanones, alkanals
	$C_5H_{12}N^+$	aliphatic amines <u>N-indicator</u>
87	$C_5H_{11}O^+$	alcohols, ethers, esters <u>O-indicator</u>
	$C_4H_7O_2^+$	esters, acids
88	$C_4H_8O_2^{+\cdot}$	ethyl esters of carboxylic acids, α-methyl-methyl esters, α-C_2-carboxylic acids
89	$C_4H_9O_2^+$	diols, glycol ethers <u>2 x O-indicator</u>
	$C_4H_9S^+$	sulfides
90	$C_7H_6^{+\cdot}$	disubstituted aromatics
91	$C_7H_7^+$	aromatics
92	$C_7H_8^{+\cdot}$	alkylbenzenes
	$C_6H_6N^+$	alkylpyridines
93	$C_6H_5O^+$	phenols, phenol derivatives
	$C_6H_7N^{+\cdot}$	anilines
	CH_2Br^+	bromides
94	$C_6H_6O^{+\cdot}$	phenol esters, phenol ethers
	$C_5H_4NO^+$	pyrryl ketones, pyridone derivatives
95	$C_5H_3O_2^+$	furyl ketones
96	$C_7H_{12}^{+\cdot}$	alicyclics
97	$C_7H_{13}^+$	cycloalkanes, alkenes
	$C_6H_9O^+$	cycloalkanones
	$C_5H_5S^+$	alkylthiophenes
98	$C_6H_{12}N^+$	N-alkylpiperidines
99	$C_7H_{15}^+$	alkanes
	$C_6H_{11}O^+$	alkanones
	$C_5H_7O_2^+$	ethylene ketals

MASS CORRELATION TABLE

Mass	Ion	Compound type
99	$H_4PO_4^+$	alkyl phosphates
104	$C_8H_8^{+\cdot}$	tetralin derivatives, phenylethyl derivatives
	$C_7H_4O^{+\cdot}$	disubstituted α-ketobenzenes
105	$C_8H_9^+$	alkyl aromatics
	$C_7H_5O^+$	benzoyl derivatives
	$C_6H_5N_2^+$	diazophenyl derivatives
111	$C_5H_3OS^+$	thiophenoyl derivatives
115	$C_9H_7^+$	aromatics
	$C_6H_{11}O_2^+$	esters
	$C_5H_7O_3^+$	diesters
119	$C_9H_{11}^+$	alkyl aromatics
	$C_8H_7O^+$	tolyl ketones
	$C_2F_5^+$	perfluoroethyl derivatives
	$C_7H_5NO^{+\cdot}$	phenylcarbamates
120	$C_7H_4O_2^+$	γ-benzpyrones, salicylic acid derivatives
	$C_8H_{10}N^+$	pyridines, anilines
121	$C_8H_9O^+$	
	$C_7H_5O_2^+$	hydroxybenzene derivatives
127	$C_{10}H_7^+$	naphthalenes
	$C_6H_7O_3^+$	unsaturated diesters
	$C_6H_6NCl^{+\cdot}$	chlorinated N-aromatics
	I^+	iodides
128	$C_{10}H_8^{+\cdot}$	naphthalenes
	$C_6H_5OCl^{+\cdot}$	chlorinated hydroxybenzene derivatives
	$HI^{+\cdot}$	iodides
130	$C_9H_8N^+$	quinolines, indoles
	$C_9H_6O^{+\cdot}$	naphthoquinones
131	$C_{10}H_{11}^+$	tetralins
	$C_5H_7S_2^+$	thioethylene ketals
	$C_3F_5^+$	perfluoroalkyl derivatives
135	$C_4H_8Br^+$	alkyl bromides

Mass	Ion	Compound type
141	$C_{11}H_9^+$	naphthalenes
142	$C_{10}H_8N^+$	quinolines
149	$C_8H_5O_3^+$	phthalates
152	$C_{12}H_8^{+\cdot}$	diphenyl aromatics
165	$C_{13}H_9^+$	diphenylmethane derivatives
167	$C_8H_7O_4^+$	phthalates

MS

ISOTOPES, DISTRIBUTION PATTERNS

Isotope Patterns of All Naturally Occurring Elements in the Periodic Table

Cl 35	Zn 64	Ru 102	Ce 140	W 184	
S 32	Cu 63	Mo 98	Ba 138	Hf 180	
Si 28	Ni 58	Zr 90	Xe 132	Lu 175	U 238
Mg 24	Fe 56	Sr 88	Te 130	Yb 174	Pb 208
Ne 20	Cr 52	Rb 85	Sb 121	Er 166	Tl 205
O 16	V 51	Kr 84	Sn 120	Dy 164	Hg 202
N 14	Ti 48	Br 79	In 115	Gd 158	Pt 195
C 12	Ca 40	Se 80	Cd 114	Eu 153	Ir 193
B 11	K 39	Ge 74	Ag 107	Sm 152	Os 192
Li 7	Ar 40	Ga 69	Pd 106	Nd 142	Re 187

H	He	Be	F	Na	Al	P	Sc	Mn	Co	As	Y	Nb	Rh	I	Cs	La	Pr	Tb	Ho	Tm	Ta	Au	Bi	Th
1	4	9	19	23	27	31	45	55	59	75	89	93	103	127	133	139	141	159	165	169	181	197	209	232

Shown under the symbol of the element is the mass of the most abundant isotope; the lightest isotope is shown at the beginning of the mass scale. The row at the bottom summarizes all the monoisotopic – or almost monoisotopic – elements.

For the exact masses and abundances see p. M60 – M85.

Mass and Relative Abundance of the Isotopes of All Naturally Occurring Elements

Isotope	Mass	Abundance	Isotope	Mass	Abundance
^{1}H	1.0078	100	^{26}Mg	25.9826	14.193
^{2}H	2.0141	0.015	^{27}Al	26.9815	100
^{3}He	3.0160	$\sim 10^{-4}$	^{28}Si	27.9769	100
^{4}He	4.0026	100	^{29}Si	28.9765	5.097
^{6}Li	6.0151	8.015	^{30}Si	29.9738	3.351
^{7}Li	7.0160	100	^{31}P	30.9738	100
^{9}Be	9.0122	100	^{32}S	31.9721	100
^{10}B	10.0129	24.394	^{33}S	32.9715	0.800
^{11}B	11.0093	100	^{34}S	33.9679	4.442
^{12}C	12.0000	100	^{36}S	35.9671	0.014
^{13}C	13.0034	1.119	^{35}Cl	34.9689	100
^{14}N	14.0031	100	^{37}Cl	36.9659	31.978
^{15}N	15.0001	0.368	^{36}Ar	35.9675	0.338
^{16}O	15.9949	100	^{38}Ar	37.9627	0.063
^{17}O	16.9991	0.037	^{40}Ar	39.9624	100
^{18}O	17.9991	0.204	^{39}K	38.9637	100
^{19}F	18.9984	100	^{40}K	39.9640	0.013
^{20}Ne	19.9924	100	^{41}K	40.9618	7.390
^{21}Ne	20.9938	0.283	^{40}Ca	39.9626	100
^{22}Ne	21.9914	9.701	^{42}Ca	41.9586	0.660
^{23}Na	22.9898	100	^{43}Ca	42.9588	0.151
^{24}Mg	23.9850	100	^{44}Ca	43.9555	2.124
^{25}Mg	24.9858	12.872	^{46}Ca	45.9537	0.003
			^{48}Ca	47.9524	0.191
			^{45}Sc	44.9559	100

Isotope	Mass	Abundance	Isotope	Mass	Abundance
^{46}Ti	45.9526	10.725	^{68}Zn	67.9249	37.983
^{47}Ti	46.9518	9.846	^{70}Zn	69.9254	1.268
^{48}Ti	47.9479	100			
^{49}Ti	48.9479	7.452	^{69}Ga	68.9257	100
^{50}Ti	49.9448	7.222	^{71}Ga	70.9248	65.563
^{50}V	49.9472	0.241	^{70}Ge	69.9243	56.158
^{51}V	50.9440	100	^{72}Ge	71.9217	75.068
			^{73}Ge	72.9234	21.237
^{50}Cr	49.9460	5.146	^{74}Ge	73.9212	100
^{52}Cr	51.9405	100	^{76}Ge	75.9214	21.237
^{53}Cr	52.9407	11.402			
^{54}Cr	53.9389	2.841	^{75}As	74.9216	100
^{55}Mn	54.9381	100	^{74}Se	73.9225	1.746
			^{76}Se	75.9192	18.105
^{54}Fe	53.9396	6.350	^{77}Se	76.9199	15.215
^{56}Fe	55.9349	100	^{78}Se	77.9174	47.210
^{57}Fe	56.9354	2.389	^{80}Se	79.9165	100
^{58}Fe	57.9333	0.360	^{82}Se	81.9167	18.446
^{59}Co	58.9332	100	^{79}Br	78.9184	100
			^{81}Br	80.9163	97.874
^{58}Ni	57.9353	100			
^{60}Ni	59.9308	38.644	^{78}Kr	77.9204	0.622
^{61}Ni	60.9311	1.754	^{80}Kr	79.9164	3.989
^{62}Ni	61.9284	5.394	^{82}Kr	81.9135	20.316
^{64}Ni	63.9280	1.592	^{83}Kr	82.9141	20.299
			^{84}Kr	83.9115	100
^{63}Cu	62.9296	100	^{86}Kr	85.9106	30.527
^{65}Cu	64.9278	44.739			
			^{85}Rb	84.9117	100
^{64}Zn	63.9292	100	^{87}Rb	86.9092	38.600
^{66}Zn	65.9261	56.883			
^{67}Zn	66.9272	8.407			

ISOTOPES, MASSES AND ABUNDANCES

Isotope	Mass	Abundance	Isotope	Mass	Abundance
^{84}Sr	83.9134	0.678	^{102}Pd	101.9051	3.513
^{86}Sr	85.9093	11.943	^{104}Pd	103.9037	40.139
^{87}Sr	86.9089	8.503	^{105}Pd	104.9046	81.339
^{88}Sr	87.9056	100	^{106}Pd	105.9032	100
			^{108}Pd	107.9039	97.731
^{89}Y	88.9059	100	^{110}Pd	109.9045	43.213
^{90}Zr	89.9048	100	^{107}Ag	106.9050	100
^{91}Zr	90.9057	21.823	^{109}Ag	108.9047	92.975
^{92}Zr	91.9051	33.249			
^{94}Zr	93.9063	33.813	^{106}Cd	105.9060	4.210
^{96}Zr	95.9084	5.441	^{108}Cd	107.9040	3.032
			^{110}Cd	109.9030	42.931
^{93}Nb	92.9065	100	^{111}Cd	110.9040	44.179
			^{112}Cd	111.9030	83.403
^{92}Mo	91.9068	66.611	^{113}Cd	112.9046	42.481
^{94}Mo	93.9052	38.015	^{114}Cd	113.9036	100
^{95}Mo	94.9059	66.106	^{116}Cd	115.9050	26.265
^{96}Mo	95.9047	69.512			
^{97}Mo	96.9060	39.781	^{113}In	112.9043	4.471
^{98}Mo	97.9055	100	^{115}In	114.9039	100
^{100}Mo	99.9075	40.496			
			^{112}Sn	111.9051	2.922
^{96}Ru	95.9074	17.431	^{114}Sn	113.9029	2.009
^{98}Ru	97.9047	5.916	^{115}Sn	114.9033	1.065
^{99}Ru	98.9057	40.240	^{116}Sn	115.9018	43.531
^{100}Ru	99.9041	39.924	^{117}Sn	116.9029	23.166
^{101}Ru	100.9052	54.002	^{118}Sn	117.9016	73.151
^{102}Ru	101.9039	100	^{119}Sn	118.9032	26.119
^{104}Ru	103.9051	58.779	^{120}Sn	119.9021	100
			^{122}Sn	121.9032	14.368
^{103}Rh	102.9041	100	^{124}Sn	123.9051	18.082

MS

ISOTOPES, MASSES AND ABUNDANCES

Isotope	Mass	Abundance	Isotope	Mass	Abundance
^{121}Sb	120.9036	100	^{139}La	138.9061	100
^{123}Sb	122.9039	74.672			
			^{136}Ce	135.9071	0.218
^{120}Te	119.9045	0.258	^{138}Ce	137.9057	0.283
^{122}Te	121.9028	7.135	^{140}Ce	139.9053	100
^{123}Te	122.9042	2.523	^{142}Ce	141.9090	12.511
^{124}Te	123.9028	13.370			
^{125}Te	124.9044	20.273	^{141}Pr	140.9074	100
^{126}Te	125.9032	54.263			
^{128}Te	127.9047	92.198	^{142}Nd	141.9075	100
^{130}Te	129.9067	100	^{143}Nd	142.9096	44.891
			^{144}Nd	143.9099	87.975
^{127}I	126.9044	100	^{145}Nd	144.9122	30.616
			^{146}Nd	145.9127	63.519
^{124}Xe	123.9061	0.357	^{148}Nd	147.9165	21.136
^{126}Xe	125.9042	0.335	^{150}Nd	149.9207	20.730
^{128}Xe	127.9035	7.136			
^{129}Xe	128.9048	98.326	^{144}Sm	143.9117	11.564
^{130}Xe	129.9035	15.173	^{147}Sm	146.9146	56.025
^{131}Xe	130.9051	78.765	^{148}Sm	147.9146	42.066
^{132}Xe	131.9042	100	^{149}Sm	148.9169	51.759
^{134}Xe	133.9054	38.825	^{150}Sm	149.9170	27.844
^{136}Xe	135.9072	32.986	^{152}Sm	151.9195	100
			^{154}Sm	153.9220	84.992
^{133}Cs	132.9051	100			
			^{151}Eu	150.9196	91.644
^{130}Ba	129.9063	0.141	^{153}Eu	152.9209	100
^{132}Ba	131.9051	0.135			
^{134}Ba	133.9043	3.377	^{152}Gd	151.9195	0.804
^{135}Ba	134.9056	9.196	^{154}Gd	153.9207	8.645
^{136}Ba	135.9044	10.899	^{155}Gd	154.9226	59.228
^{137}Ba	136.9056	15.797	^{156}Gd	155.9221	82.308
^{138}Ba	137.9050	100	^{157}Gd	156.9239	63.048
			^{158}Gd	157.9241	100
^{138}La	137.9068	0.089	^{160}Gd	159.9271	88.058

Isotope	Mass	Abundance	Isotope	Mass	Abundance
^{159}Tb	158.9250	100	^{178}Hf	177.9439	77.015
			^{179}Hf	178.9460	39.018
^{156}Dy	155.9238	0.186	^{180}Hf	179.9468	100
^{158}Dy	157.9240	0.320			
^{160}Dy	159.9248	8.141	^{180}Ta	179.9475	0.012
^{161}Dy	160.9266	66.998	^{181}Ta	180.9491	100
^{162}Dy	161.9265	90.596			
^{163}Dy	162.9284	88.609	^{180}W	179.9470	0.441
^{164}Dy	163.9288	100	^{182}W	181.9483	86.194
			^{183}W	182.9503	46.997
^{165}Ho	164.9303	100	^{184}W	183.9510	100
			^{186}W	185.9543	92.722
^{162}Er	161.9288	0.407			
^{164}Er	163.9293	4.669	^{185}Re	184.9530	58.907
^{166}Er	165.9304	100	^{187}Re	186.9560	100
^{167}Er	166.9321	68.662			
^{168}Er	167.9324	81.024	^{184}Os	183.9526	0.044
^{170}Er	169.9355	44.538	^{186}Os	185.9539	3.878
			^{187}Os	186.9560	4.000
^{169}Tm	168.9344	100	^{188}Os	187.9560	32.439
			^{189}Os	188.9583	39.268
^{168}Yb	167.9339	0.424	^{190}Os	189.9586	64.390
^{170}Yb	169.9349	9.516	^{192}Os	191.9614	100
^{171}Yb	170.9365	44.943			
^{172}Yb	171.9366	68.530	^{191}Ir	190.9609	59.489
^{173}Yb	172.9383	50.660	^{193}Ir	192.9633	100
^{174}Yb	173.9390	100			
^{176}Yb	175.9427	39.981	^{190}Pt	189.9560	0.038
			^{192}Pt	191.9614	2.308
^{175}Lu	174.9409	100	^{194}Pt	193.9628	97.337
^{176}Lu	175.9427	2.659	^{195}Pt	194.9648	100
			^{196}Pt	195.9650	74.852
^{174}Hf	173.9403	0.511	^{198}Pt	197.9675	21.331
^{176}Hf	175.9417	14.756			
^{177}Hf	176.9435	52.497	^{197}Au	196.9666	100

MS

Isotope	Mass	Abundance
^{196}Hg	195.9658	0.490
^{198}Hg	197.9668	33.624
^{199}Hg	198.9683	56.510
^{200}Hg	199.9683	77.617
^{201}Hg	200.9703	44.362
^{202}Hg	201.9706	100
^{204}Hg	203.9735	22.987
^{203}Tl	202.9723	41.844
^{205}Tl	204.9745	100
^{204}Pb	203.9731	2.830
^{206}Pb	205.9745	45.124
^{207}Pb	206.9759	43.212
^{208}Pb	207.9766	100
^{209}Bi	208.9804	100
^{232}Th	232.0382	100
^{234}U	234.0409	0.006
^{235}U	235.0439	0.726
^{238}U	238.0508	100

Calculation of Isotope Distributions

The characteristic abundance patterns resulting from the combination
of more than one polyisotopic element can be calculated using the
following scheme:

B ↓ \ A →		m_{a_1}	$m_{a_1}+1$	$m_{a_1}+2$	$m_{a_1}+3$	$m_{a_1}+4$	$m_{a_1}+5\ldots$
		$a_1=1$	a_2	a_3	a_4		
m_{b_1}	$b_1=1$	$1\cdot1$	$1\cdot a_2$	$1\cdot a_3$	$1\cdot a_4$		
$m_{b_1}+1$	b_2		$b_2\cdot1$	$b_2\cdot a_2$	$b_2\cdot a_3$	$b_2\cdot a_4$	
$m_{b_1}+2$	b_3			$b_3\cdot1$	$b_3\cdot a_2$	$b_3\cdot a_3$	$b_3\cdot a_4$
$m_{b_1}+3$ $\vdots$		1	b_2+a_2	$(b_2\cdot a_2)$ $+a_3$ $+b_3$	$(b_2\cdot a_3)$ $+(b_3\cdot a_2)$ $+a_4$	$(b_2\cdot a_4)$ $+(b_3\cdot a_3)$	$b_3\cdot a_4$
	$m_{a_1}+m_{b_1}$	$+1$	$+2$	$+3$	$+4$	$+5$

An element A consisting of the isotopes of mass m_{a_1}, m_{a_2}, m_{a_3}, m_{a_4}
of the natural abundances a_1, a_2, a_3, a_4 are combined with element B
with isotopes of mass m_{b_1}, m_{b_2}, m_{b_3} of the natural abundance b_1, b_2,
b_3.

The relative abundances of the lightest isotopes m_{a_1} and m_{b_1} are
normalized to make their abundances $a_1 = b_1 = 1$. The normalized
abundances of element A are entered according to their increasing
mass in the first horizontal row and those of element B in the first

vertical column. Each entry in a row is then multiplied with that of the column, moving from left to right. The sum of each column represents the abundance distribution of A and B by increasing mass.

The resulting distribution can be entered as a "super element" to be combined with additional elements or "super elements" using the same computational scheme.

Example: What is the isotope pattern of $ZnBr_2$?
Zn consists of isotopes of mass 64, 66, 67, 68 and 70 (see p. M65 and M55) of the abundance distribution (rounded off) 1 : 0.57 : 0.08 : 0.38 : 0.01. Br_2 is entered as a "super element" consisting of the isotopes of mass 158, 160 and 162 (see p. M105 and M100) and an abundance distribution (rounded off) 1 : 2 : 1.

Zn →		64	65	66	67	68	69	70	71	72	73	74
Br_2 ↓		1		0.57	0.08	0.38		0.01				
158	1	1		0.57	0.08	0.38		0.01				
159												
160	2			2		1.14	0.16	0.76		0.02		
161												
162	1					1		0.57	0.08	0.38		0.01
		1		2.57	0.08	2.52	0.16	1.34	0.08	0.40		0.01

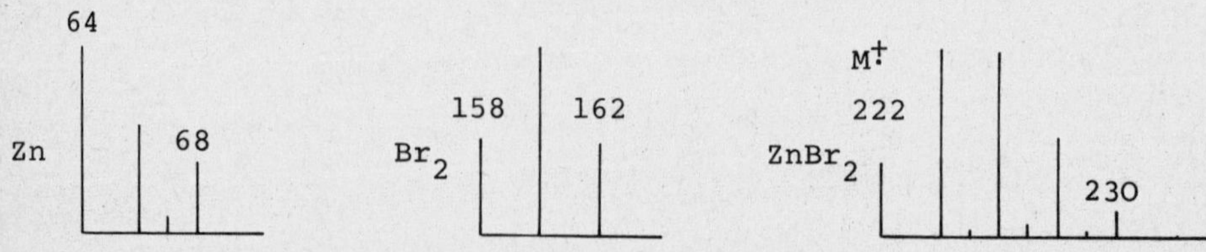

Isotope Patterns of Various Combinations of Cl and Br

<u>Note:</u> The signals are separated by 2 mass units.

For exact abundances see p. M105, M110.

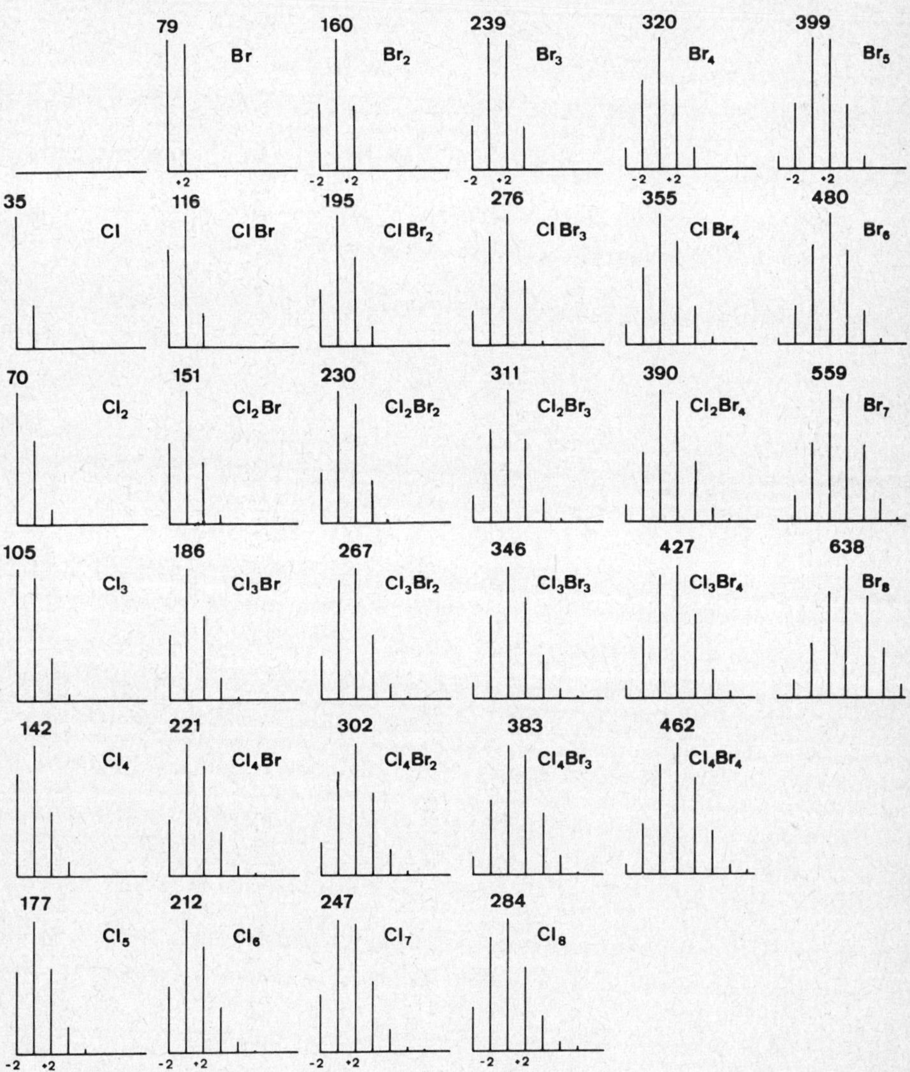

HALOGENS, ISOTOPE DISTRIBUTIONS

Isotopic Abundances of Various Combinations of Chlorine and Bromine

Halogens	Mass	Rel. Abundance	Halogens	Mass	Rel. Abundance
Cl_1	35	100	Br_1	79	100
	37	31.98		81	97.88
Cl_2	70	100	Br_2	158	51.09
	72	63.96		160	100
	74	10.23		162	48.93
Cl_3	105	100	Br_3	237	34.05
	107	95.93		239	100
	109	30.67		241	97.89
	111	3.27		243	31.94
Cl_4	140	77.96	Br_4	316	17.40
	142	100		318	68.09
	144	47.82		320	100
	146	10.19		322	65.26
	148	0.82		324	15.96
Cl_5	175	62.53	Br_5	395	10.43
	177	100		397	51.09
	179	63.94		399	100
	181	20.45		401	97.94
	183	3.28		403	47.89
	185	0.21		405	9.38
Cl_6	210	52.12	Br_6	474	5.32
	212	100		476	31.26
	214	79.95		478	76.62
	216	34.08		480	100
	218	8.21		482	73.38
	220	1.05		484	28.73
	222	0.06		486	4.68
Cl_1Br_1	114	76.70	Cl_1Br_4	351	14.26
	116	100		353	60.41
	118	24.46		355	100
				357	79.93
Cl_1Br_2	193	43.83		359	30.39
	195	100		361	4.25
	197	69.83			
	199	13.66	Cl_1Br_5	430	8.02
				432	41.85
Cl_1Br_3	272	26.15		434	89.50
	274	85.22		436	100
	276	100		438	61.10
	278	48.90		440	19.12
	280	7.86		442	2.35

Halogens	Mass	Rel. Abundance	Halogens	Mass	Rel. Abundance
Cl_2Br_1	149	61.35	Cl_4Br_1	219	43.79
	151	100		221	100
	153	45.67		223	83.86
	155	6.38		225	33.42
				227	6.93
Cl_2Br_2	228	38.35		229	0.48
	230	100			
	232	89.63	Cl_4Br_2	298	24.14
	234	31.89		300	78.63
	236	3.90		302	100
				304	63.54
Cl_2Br_3	307	20.49		306	21.54
	309	73.38		308	3.73
	311	100		310	0.26
	313	63.78			
	315	18.71	Cl_4Br_3	377	13.63
	317	2.03		379	57.78
				381	100
Cl_2Br_4	386	11.92		383	91.19
	388	54.36		385	47.31
	390	100		387	14.03
	392	94.03		389	2.22
	394	47.21		391	0.13
	396	11.82			
	398	1.15	Cl_5Br_1	254	37.60
				256	98.11
Cl_3Br_1	184	51.12		258	100
	186	100		260	52.18
	188	65.22		262	14.89
	190	17.73		264	2.22
	192	1.74		266	0.12
Cl_3Br_2	263	31.35	Cl_5Br_2	333	19.19
	265	92.01		335	68.85
	267	100		337	100
	269	50.01		339	76.56
	271	11.70		341	33.64
	273	1.03		343	8.56
				345	1.17
Cl_3Br_3	342	16.50		347	0.06
	344	64.58			
	346	100			
	348	77.78			
	350	31.90			
	352	6.58			
	354	0.54			

INDICATIONS OF STRUCTURAL TYPE

Indications of Structural Type

Certain sequences of abundance maxima in the lower mass range and the masses of unique signals are often characteristic of a particular compound type. The abundance distribution of such ion series is

Mass values / Compound types	Alkanes	Alkenes, Mono-cycloalkanes	Alkynes, Dienes, Cycloalkenes	Poly-cyclic alicycles	Alkanones, Alkanals	Alcohols, Alkyl ethers	Cyclic alcohols	Cyclo-alkanones	Aliphat.. acids, Esters, Lactones	Alkyl-amines
12 + $(14)_n$ m/z 26,40,54,68, 82,96,110...		C_nH_{2n-2}	C_nH_{2n-2}	C_nH_{2n-2}				C_nH_{2n-2}		
13 + $(14)_n$ m/z 27,41,55,69, 83,97,111...	C_nH_{2n-1}	C_nH_{2n-1}	C_nH_{2n-1}	C_nH_{2n-1}		C_nH_{2n-1}	C_nH_{2n-1}	$\frac{C_nH_{2n-3}O}{C_nH_{2n-1}}$	C_nH_{2n-1}	
14 + $(14)_n$ m/z 28,42,56,70, 84,98,112...	C_nH_{2n}	C_nH_{2n}		C_nH_{2n}		C_nH_{2n}			$C_nH_{2n-2}O$	
15 + $(14)_n$ m/z 29,43,57,71, 85,99,113...	C_nH_{2n+1}	C_nH_{2n+1}	C_nH_{2n+1}	C_nH_{2n+1}	C_nH_{2n+1} $C_nH_{2n-1}O$		$C_nH_{2n-1}O$			
16 + $(14)_n$ m/z 30,44,58,72, 86,100,114...					$C_nH_{2n}O$					$C_nH_{2n+2}N$
17 + $(14)_n$ m/z 31,45,59,73, 87,101,115...						$C_nH_{2n+1}O$			$C_nH_{2n+1}O$ $C_nH_{2n-1}O_2$	
18 + $(14)_n$ m/z 32,46,60,74, 88,102,116...									$C_nH_{2n}O_2$	
19 + $(14)_n$ m/z 33,47,61,75, 89,103,117...						$C_nH_{2n+3}O$			$C_nH_{2n+1}O_2$	
20 + $(14)_n$ m/z 34,48,62,76, 90,104,118...										
21 + $(14)_n$ m/z 35,49,63,77, 91,105,119...										
22 + $(14)_n$ m/z 36,50,64,78, 92,106,120...				C_nH_{2n-6}						
23 + $(14)_n$ m/z 37,51,65,79, 93,107,121...				C_nH_{2n-5}						
24 + $(14)_n$ m/z 38,52,66,80, 94,108,122...				C_nH_{2n-4}						
25 + $(14)_n$ m/z 39,53,67,81, 95,109,123...			C_nH_{2n-3}	C_nH_{2n-3}						
11,12,13,14+$(13)_n$ m/z 39,52±1,64±1, 78±1,91±1,...										

in general smooth. Abrupt abundance changes (maxima and minima) are therefore always of structural significance. The ion or ion series which is most indicative of a particular compound type is underlined in the table.

Aliphat. amides	Thiols, Sulfides	Glycols, Glycol ethers	Alkyl chlorides	Acid chlorides	Phenols, Aryl ethers	Aryl ketones	Aromat. hydrocarbons	Alkyl-benzenes	Alkyl-anilines	Compound types / Mass values
			C_nH_{2n+1}							$12 + (14)_n$ m/z 26,40,54,68, 82,96,110...
	C_nH_{2n-1}	C_nH_{2n-1}	C_nH_{2n-1}							$13 + (14)_n$ m/z 27,41,55,69 83,97,111...
	C_nH_{2n}		C_nH_{2n}							$14 + (14)_n$ m/z 28,42,56,70, 84,98,112...
		C_nH_{2n+1}		C_nH_{2n+1} $C_nH_{2n-1}O$						$15 + (14)_n$ m/z 29,43,57,71, 85,99,113...
$\underline{C_nH_{2n+2}N}$ $C_nH_{2n}NO$										$16 + (14)_n$ m/z 30,44,58,72, 86,100,114...
		$C_nH_{2n+1}O$								$17 + (14)_n$ m/z 31,45,59,73, 87,101,115...
										$18 + (14)_n$ m/z 32,46,60,74, 88,102,116...
	$\underline{C_nH_{2n+1}S}$	$\underline{C_nH_{2n+1}O_2}$								$19 + (14)_n$ m/z 33,47,61,75, 89,103,117...
	$C_nH_{2n+2}S$	$C_nH_{2n+2}O_2$						$C_8H_8 +$ C_nH_{2n}		$20 + (14)_n$ m/z 34,48,62,76, 90,104,118...
			$\underline{C_nH_{2n}Cl}$	COCl		$C_7H_5O +$ C_nH_{2n}		$C_7H_7 +$ C_nH_{2n}		$21 + (14)_n$ m/z 35,49,63,77, 91,105,119...
									$C_6H_6N +$ C_nH_{2n}	$22 + (14)_n$ m/z 36,50,64,78, 92,106,120...
					$C_nH_{n-1}O$ $+ C_nH_{2n}$					$23 + (14)_n$ m/z 37,51,65,79, 93,107,121...
					C_nH_nO					$24 + (14)_n$ m/z 38,52,66,80, 94,108,122...
										$25 + (14)_n$ m/z 39,53,67,81, 95,109,123...
					$C_nH_{n\pm1}$	$C_nH_{n\pm1}$	$\underline{C_nH_{n\pm1}}$	$C_nH_{2n\pm1}$		$11,12,13,14+(13)_n$ m/z 39,52±1,64±1, 78±1,91±1,...

MS

INDICATIONS OF THE PRESENCE
OF HETEROATOMS

Indications of the Presence of Heteroatoms

In unit-resolution mass spectra one often observes characteristic
abundances of ions due to isotope patterns, specific masses of fragment ions and characteristic mass differences (Δm) between the molecular ion ($M^{+\cdot}$) and fragment ions (F^{+}) or between pairs of fragment
ions. It should be noted that the mass differences between fragment
ions can be used reliably only if their relationship is supported by
the observation of the corresponding signal ("metastable peak", m^{*})
due to this unimolecular decomposition.

Indication of O: Δm 17 from $M^{+\cdot}$, in N-free compounds

Δm 18 from $M^{+\cdot}$

Δm 18 from F^{+} (m^{*}), particularly in aliphatic compounds

Δm 28, 29 from $M^{+\cdot}$ for aromatic compounds

Δm 28 from F^{+} (m^{*}) for aromatic compounds

m/z 15, relatively abundant

m/z 19

m/z 31, 45, 59, 73 ... + $(14)_n$

m/z 32, 46, 60, 74 ... + $(14)_n$

m/z 33, 47, 61, 75 ... + $(14)_n$ for 2 x O, in the
absence of S

m/z 69 for aromatic compounds meta-disubstituted
by oxygen

Indication of N: $M^{+\cdot}$ odd-numbered

large number of even-numbered fragment ions

Δm 17 from $M^{+\cdot}$ or F^{+} (m^{*}), in O-free compounds

Δm 27 from $M^{+\cdot}$ or F^{+} (m^{*}) for aromatic compounds or
nitriles

Δm 30, 46 for nitro compounds

m/z 30, 44, 58, 72 ... + $(14)_n$ for aliphatic compounds

M125

Indication of S: Isotope peak $M^{+\cdot} + 2 \geqslant 5\%$ $M^{+\cdot}$

Δm 33, 34, 47, 48, 64, 65 from $M^{+\cdot}$

Δm 34, 48, 64 from F^{+} (m^*)

m/z 33, 34, 35

m/z 45, in O-free compounds

m/z 47, 61, 75, 89 ... + $(14)_n$, unless compound contains 2 oxygens

m/z 48, 64 for S-oxides

Indication of F: Δm 19, 20, 50 from $M^{+\cdot}$

Δm 20 from F^{+} (m^*)

m/z 20

m/z 57 without m/z 55 in aromatics

Indication of Cl: Isotope peak $[M+2]^{+\cdot} \geqslant 33\%$ $M^{+\cdot}$

Δm 35, 36 from $M^{+\cdot}$

Δm 36 from F^{+} (m^*)

m/z 35/37, 36/38, 49/51

Indication of Br: Isotope peak $[M+2]^{+\cdot} \geqslant 98\%$ $M^{+\cdot}$

Δm 79, 80 from $M^{+\cdot}$

Δm 80 from F^{+} (m^*)

m/z 79/81, 80/82

Indication of I: Isotope peak $M^{+\cdot} + 1$ of very low abundance at relatively high mass

Δm 127 from $M^{+\cdot}$

Δm 127, 128 from F^{+} (m^*)

m/z 127, 128, 254

Indication of P: m/z 47, in compounds free of S or two oxygens

m/z 99 without isotope peak at m/z 100 in alkyl-phosphates

Rules for the Determination of Relative Molecular Weight (M_r)

- The molecular ion ($M^{+\cdot}$) is defined as that ion the mass of which is the sum of all elements present in the molecule, using the most abundant isotope in each case.

- $M^{+\cdot}$ is always accompanied by isotope peaks. Their relative abundance depends on the number and kind of the elements present and their natural isotopic distribution. The abundance of $[M^{+\cdot} + 1]$ indicates the maximum number of carbon atoms (C_{max}) according to the following relationship:

$$C_{max} = \frac{[M^{+\cdot} + 1] \times 100}{[M^{+\cdot}] \times 1.1} \ .$$

$[M^{+\cdot} + 2]$ and higher masses indicate the number and kind of elements that have a relatively abundant isotope two mass units heavier (such as S, Si, Cl, Br).

- $M^{+\cdot}$ is always an even number if the molecule contains only elements for which the atomic mass and valence are both even-numbered or both odd-numbered (such as H, C, O, S, Si, P, F, Cl, Br, I). In the presence of other elements $M^{+\cdot}$ becomes an odd number unless these elements are present in an even number (this holds for N, ^{13}C, ^{2}H).

- $M^{+\cdot}$ can only form fragment ions of a mass which differs from that of the molecular ion by a chemically logical value (Δm). In this context chemically illogical differences are: $\Delta m = 3$ (in the absence of $\Delta m = 1$) to $\Delta m = 14$, $\Delta m = 21$ (in the absence of $\Delta m = 1$) to $\Delta m = 24$, $\Delta m = 37$, 38 and all Δm less than the mass of an element of characteristic isotope pattern in cases where the same isotope pattern is not retained in the fragment ion.

- $M^{+\cdot}$ of a compound must contain all elements (and the maximum number of each) that are shown to be present in the fragment ions.

- $M^{+\cdot}$ is the ion with lowest appearance potential.

- $M^{+\cdot}$ exhibit the same effusion rate as can be determined from the fragment ions (this requires that the sample flows into the ion source through a molecular leak).

- The abundance of $M^{+\cdot}$ ($[M^{+\cdot}]$) is proportional to the sample pressure in the ion source.

- In the absence of a signal for $M^{+\cdot}$ the molecular weight must have a value which shows a logical and reasonable mass difference Δm to all observed fragment ions.

- MH^+ ($M^{+\cdot} + 1$) is often observed in the mass spectra of polar compounds.

- The abundance of MH^+ changes in proportion to the square of the sample pressure in the ion source.

- For chemical ionization, field ionization, field desorption and fast atom bombardment MH^+ (or $M^{+\cdot} - H\cdot$) dominates over $M^{+\cdot}$.

MS

Signals Due to a Unimolecular Fragmentation Step ("Metastable
Peaks", m^*)

Mass spectra recorded in magnetic sector instruments often exhibit
weak, broad peaks which are due to the fragmentation of ions after
they have left the ion source. They are generally called "metastable
peaks", m^*, and are very useful in the interpretation process. If
the spectrum is recorded in the conventional way one observes almost
exclusively products resulting from the processes of the type
$m_1^+ \rightarrow m_2^+ + (m_1 - m_2)$ and $m_1^{++} \rightarrow m_2^{++} + (m_1 - m_2)$. The ion m_2^+ is
recorded at the apparent mass m^* which is related to m_1^+ according
to the equation

$$m^* = \frac{m_2^{\,2}}{m_1}$$

and thus defines the origin of m_2^+. The assignment of m_1 and m_2 to
a particular m^* is done by trial and error. The calculation is a
simple operation on any pocket calculator based on the rearranged
equation $m_2^+ = \sqrt{m^* \times m_1}$. One stores m^*, multiplies it consecutively
with the mass of each possible parent ion (m_1) and takes the square
root of the product. The result is the corresponding daughter ion.
If this is present in the spectrum, the two ions are a possibility
for a unimolecular fragmentation step, as long as the mass difference
between the two makes chemical sense. As a general approximation one
can use the fact that on a linear mass scale and for small mass differences the distances between m_1^+ and m_2^+ are approximately the same
as those between m_2^+ and m^*.

In bar graph representations of mass spectra the position of metastable peaks is often indicated as small arrows along the x-axis;
in fragmentation schemes those processes that are substantiated by
a metastable peak are indicated by an asterisk beneath the arrow
representing that process ($m_1 \xrightarrow{*} m_2$); and for mass spectra in tabular
form they are listed as $m^* = m_1^+ \rightarrow m_2^+ + (m_1 - m_2)$ or appropriate
shorter versions (see the examples on p. M150).

Example:

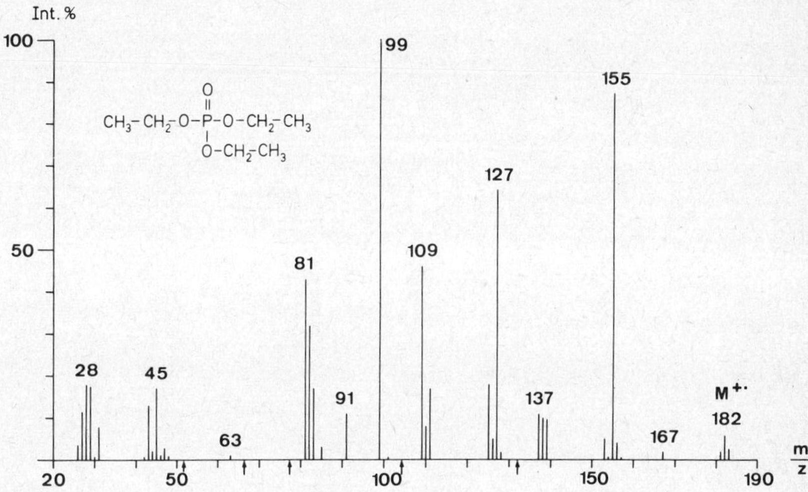

m^*	m_1^+		m_2^+		$(m_1 - m_2)$
132.0	182	⟶	155	+	27
104.1	155	⟶	127	+	28
77.2	127	⟶	99	+	28
66.3	99	⟶	81	+	18
51.7	127	⟶	81	+	46

MS

SOLVENT SPECTRA

The label [50] along the intensity scale indicates that it ends at
50% relative intensity and is subdivided in 5% steps. In those cases
the abundance of the base peak has to be doubled to bring it to 100%.

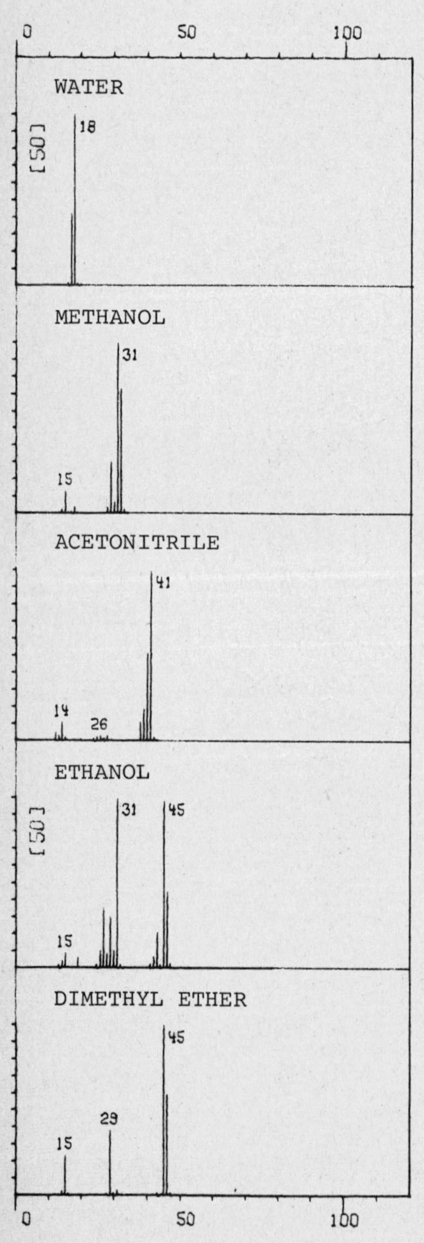

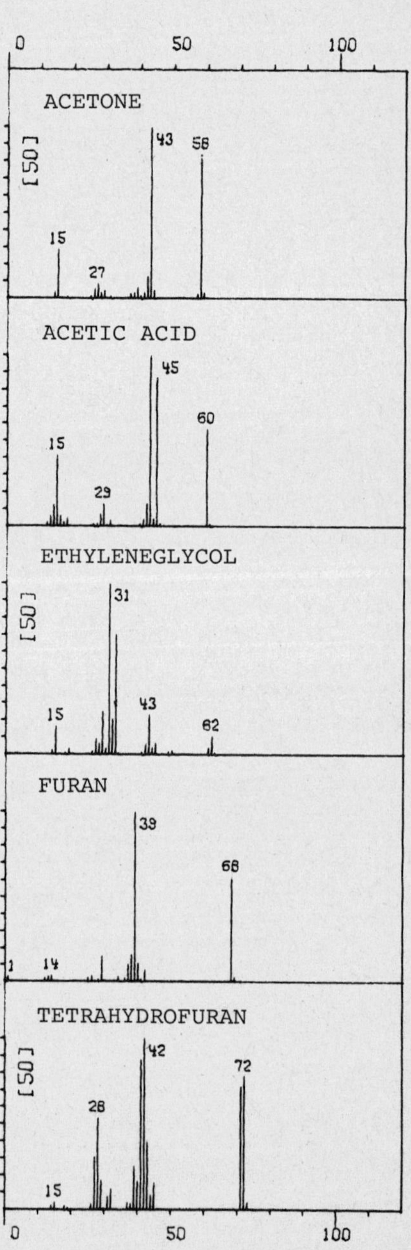

M155

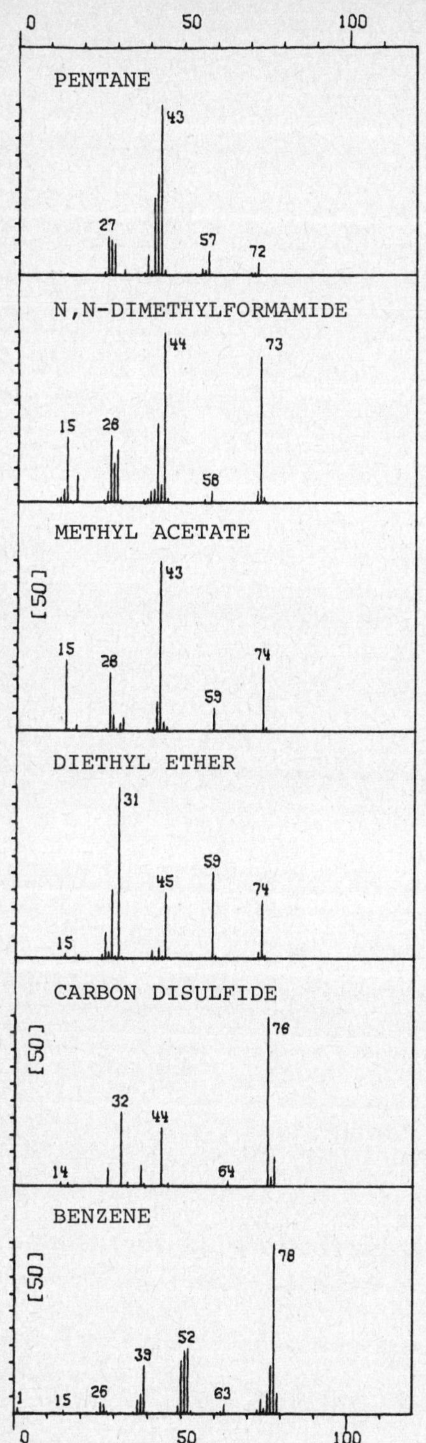
PENTANE
43
27
57
72
N,N-DIMETHYLFORMAMIDE
44
73
15
28
58
METHYL ACETATE
[50]
43
15
28
59
74
DIETHYL ETHER
31
59
45
74
15
CARBON DISULFIDE
[50]
76
32
44
14
64
BENZENE
[50]
78
52
39
1
15
26
63

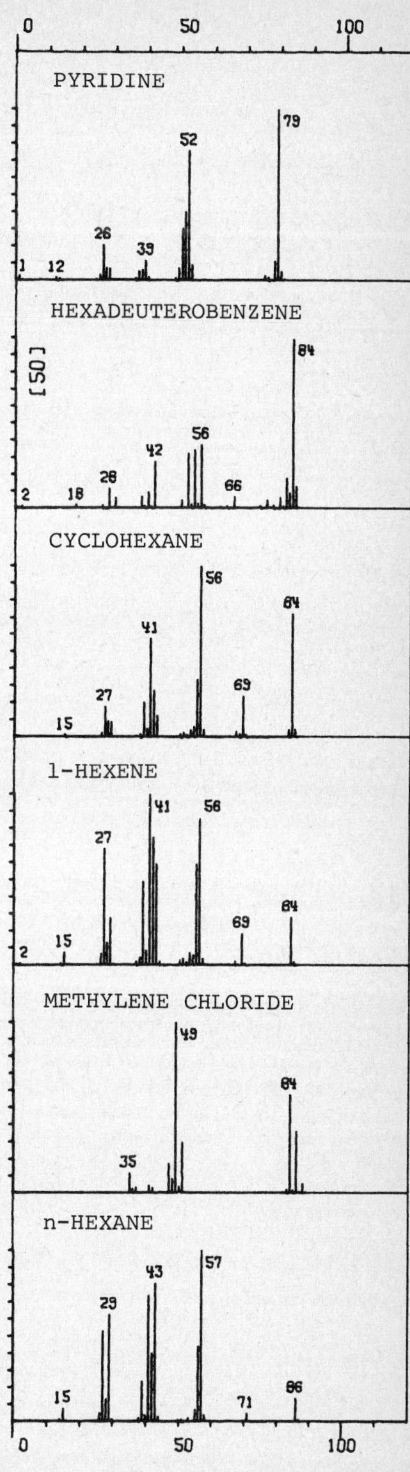
PYRIDINE
79
52
26
39
1
12
HEXADEUTEROBENZENE
[50]
84
56
42
28
2
18
66
CYCLOHEXANE
56
84
41
27
69
15
1-HEXENE
41
56
27
2
15
69
84
METHYLENE CHLORIDE
49
84
35
n-HEXANE
43
57
29
15
71
86

MS

SOLVENT SPECTRA

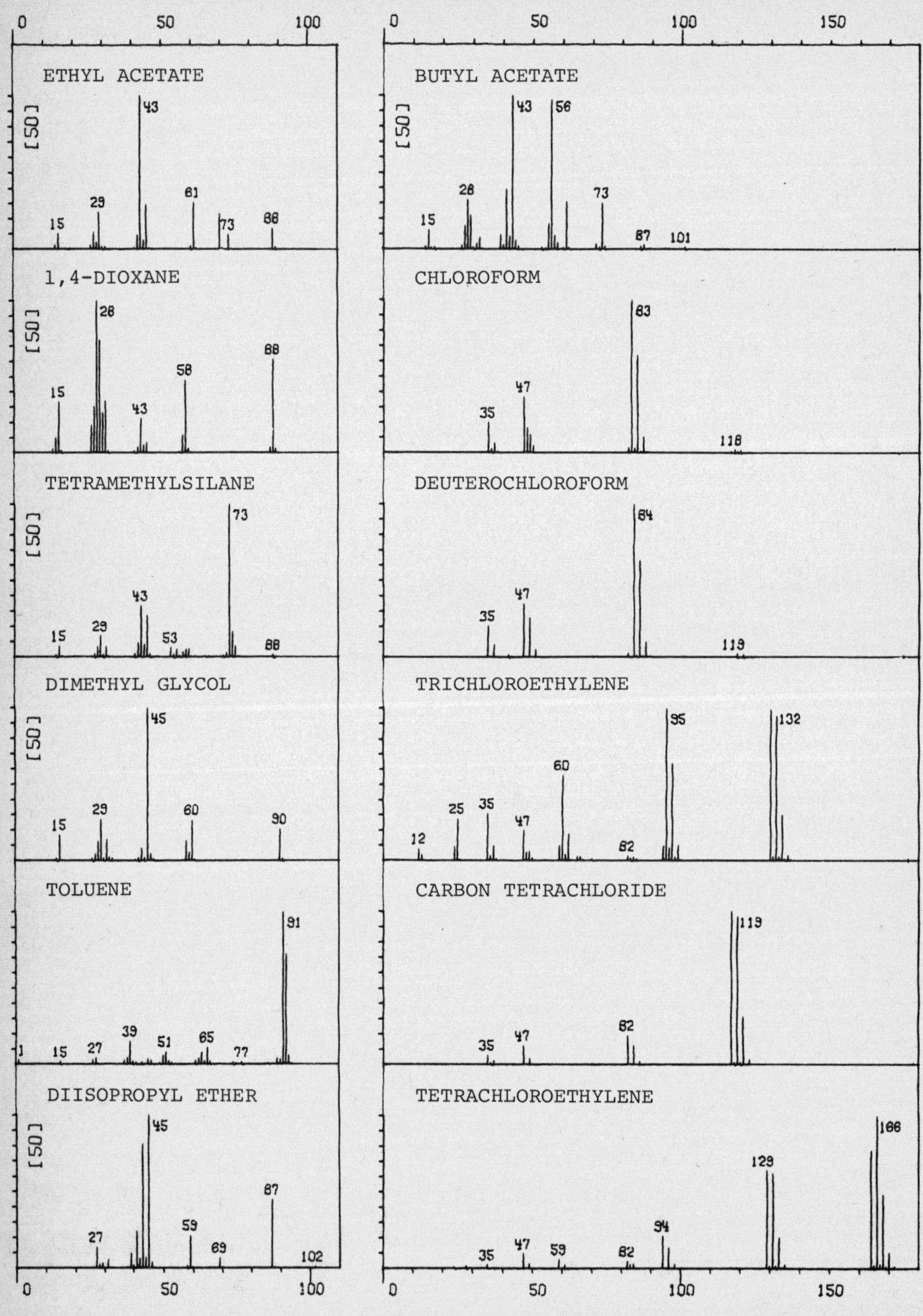

M165

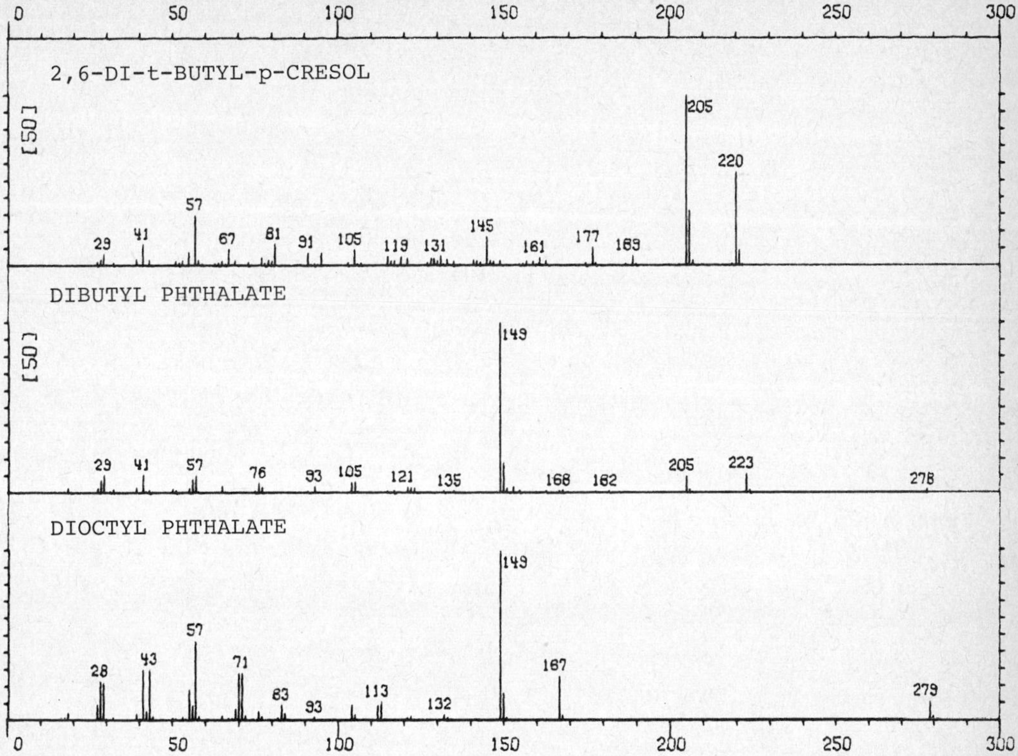

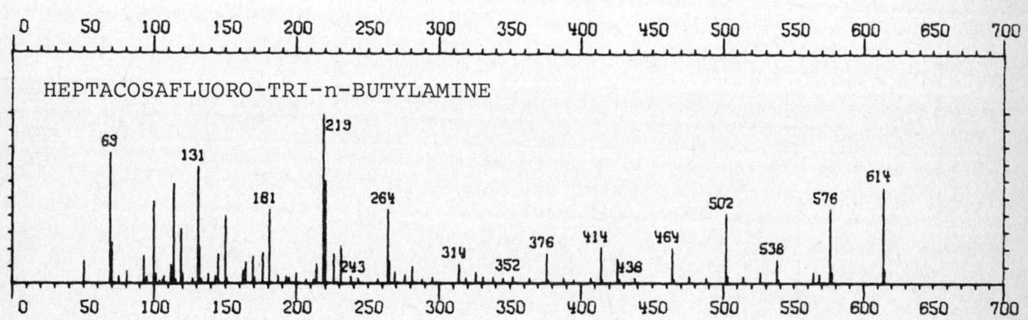

0
50
100
150
200
250
300
2,6-DI-t-BUTYL-p-CRESOL
[50]
29
41
57
67
81
91
105
119
131
145
161
177
189
205
220
DIBUTYL PHTHALATE
[50]
29
41
57
76
93
105
121
135
149
168
182
205
223
278
DIOCTYL PHTHALATE
28
43
57
71
83
93
113
132
149
167
279

0
50
100
150
200
250
300
350
400
450
500
550
600
650
700
HEPTACOSAFLUORO-TRI-n-BUTYLAMINE
69
131
181
219
243
264
314
352
376
414
438
464
502
538
576
614

Correlation Between Wavelength of Absorbed Radiation and Observed
Color

Absorbed Light		Observed (transmitted) color
Wavelength nm	Corresponding color	
400	violet	yellow-green
425	indigo blue	yellow
450	blue	orange
490	blue-green	red
510	green	purple
530	yellow-green	violet
550	yellow	indigo blue
590	orange	blue
640	red	blue-green
730	purple	green

UV/VIS

SIMPLE CHROMOPHORES

UV/VIS-Absorption of Simple Chromophores

Chromophore	Compound	Transition	λ_{max} [nm]	ε_{max}	Solvent
C–C	CH_3–CH_3	$\sigma \rightarrow \sigma^*$	135	strong	gas
C–H	CH_4	$\sigma \rightarrow \sigma^*$	122	strong	gas
C–O	CH_3OH	$n \rightarrow \sigma^*$	177	200	hexane
	CH_3–O–CH_3	$n \rightarrow \sigma^*$	184	2500	gas
C–N	$(C_2H_5)_2NH$	$n \rightarrow \sigma^*$	193	2500	hexane
	$(CH_3)_3N$	$n \rightarrow \sigma^*$	199	4000	hexane
C–S	CH_3–SH	$n \rightarrow \sigma^*$	195	1800	gas
		$n \rightarrow \sigma^*$	235	180	gas
	C_2H_5–S–C_2H_5	$n \rightarrow \sigma^*$	194	4500	gas
		$n \rightarrow \sigma^*$	225	1800	gas
S–S	C_2H_5–S–S–C_2H_5	$n \rightarrow \sigma^*$	194	5500	hexane
		$n \rightarrow \sigma^*$	250	380	
C–Cl	CH_3Cl	$n \rightarrow \sigma^*$	173	200	hexane
C–Br	n-C_3H_7Br	$n \rightarrow \sigma^*$	208	300	hexane
C–I	CH_3I	$n \rightarrow \sigma^*$	259	400	hexane
C=C	CH_2=CH_2	$\pi \rightarrow \pi^*$	162.5	15000	heptane
	$(CH_3)_2C$=$C(CH_3)_2$	$\pi \rightarrow \pi^*$	196.5	11500	heptane
C=O	$(CH_3)_2$–C=O	$n \rightarrow \sigma^*$	166	16000	gas
		$\pi \rightarrow \pi^*$	189	900	hexane
		$n \rightarrow \pi^*$	279	15	hexane
	CH_3–$\overset{O}{\overset{\|}{C}}$–OH	$n \rightarrow \pi^*$	200	50	gas
	CH_3–$\overset{O}{\overset{\|}{C}}$–$OC_2H_5$	$n \rightarrow \pi^*$	210	50	gas
	CH_3–$\overset{O}{\overset{\|}{C}}$–ONa	$n \rightarrow \pi^*$	210	150	water
	CH_3–$\overset{O}{\overset{\|}{C}}$–$NH_2$	$n \rightarrow \pi^*$	220	63	water
	$\begin{matrix} CH_2-C{\overset{\nearrow O}{\searrow}} \\ \| \qquad NH \\ CH_2-C{\overset{\searrow}{\nwarrow}} \\ \qquad\quad O \end{matrix}$		191	15200	aceto-nitrile

Chromophore	Compound	Transition	λ_{max} [nm]	ε_{max}	Solvent
C=N	$H_2N-\overset{\overset{NH}{\|\|}}{C}-NH_2 \cdot HCl$		265	15	water
	$(CH_3)_2C=NOH$		193	2000	ethanol
	$(CH_3)_2-C=NONa$		265	200	ethanol
N=N	$CH_3-N=N-CH_3$		340	16	ethanol
N=O	$(CH_3)_3C-NO$		300	100	ether
			665	20	
	$(CH_3)_3C-NO_2$		276	27	ethanol
	$n-C_4H_9-O-NO$		218	1050	
			313–384	20–40	ethanol
	$C_2H_5-O-NO_2$		260	15	ethanol
C=S	$CH_3-\overset{\overset{S}{\|\|}}{C}-CH_3$		460	weak	
	cyclohexanethione (=S)		495	weak	ethanol
C≡C	$HC≡CH$		173	6000	gas
	$n-C_5H_{11}-C≡C-CH_3$		177.5	10000	hexane
			196	2000	
			222.5	160	
C≡N	$CH_3-C≡N$		<190		
X=C=Y	$CH_2=C=CH_2$		170	4000	
			227	630	
	$(C_2H_5)_2C=C=O$		227	360	
			375	20	
	$C_2H_5-N=C=N-C_2H_5$		230	4000	
			270	25	
	$C_2H_5-N=C=S$		250	1200	hexane

UV/VIS

α,β-UNSATURATED CARBONYL COMPOUNDS

<u>UV-Absorption of α,β-Unsaturated Carbonyl Compounds</u>

(extended Woodward rules to estimate the position of the $\pi \to \pi^*$ transition)

$$\overset{\delta}{-C}=\overset{}{C}-\overset{\beta}{C}=\overset{}{C}-\overset{X}{C}=O$$
$$\quad\quad\overset{\gamma}{}\quad\quad\overset{\alpha}{}$$

Parent system:

	X : alkyl	215 nm
	X : H	207 nm
	X : OH, O-alkyl	193 nm

	215 nm

	202 nm

Increments: for each additional conjugated double bond   +30 nm

for each exocyclic double bond (C=C)

     + 5 nm

for each homoannular diene system
(homoannular arrangement of double bonds)

     +39 nm

For each substituent at the π-electron system	(Increment in nm)			
	α	β	γ	δ and beyond
C-substituent	10	12	18	18 18
OH	35	30		50 50
$OCOCH_3$	6	6	6	6 6
O-alkyl	35	30	17	31 31
S-alkyl		85		
Cl	15	12		
Br	25	30		
$N(alkyl)_2$		95		

Solvent corrections:

water	+ 8 nm
ethanol, methanol	0 nm
chloroform	- 1 nm
dioxane	- 5 nm
diethyl ether	- 7 nm
hexane, cyclohexane	-11 nm

To estimate the position of the absorption maximum in a particular solvent the appropriate increments are added to the base value for the parent system. For cross-conjugated systems the value for the chromophore absorbing at the longest wavelength should be calculated.

Example:

Estimating the absorption maximum for

in ethanolic solution.

	nm
base value	215
2 additional conjugated double bonds	60
exocyclic double bond	5
homoannular diene system	39
C-substituent in β	12
3 additional C-substituents	54
solvent correction	0
estimated:	385 nm (ethanol)
determined:	388 nm (ethanol)

UV/VIS

UV-Absorption of Dienes and Polyenes

(Woodward-Fieser rules to estimate the position of the $\pi \rightarrow \pi^*$
transition)

Parent system:

acyclic	217 nm	(and in non-con-densed ring systems)
heteroannular	214 nm	
homoannular	253 nm	

Increments:

for each additional conjugated double bond   +30 nm

for each exocyclic double bond (C=C)   + 5 nm

for each substituent:

C-substituent	+ 5 nm
$OCOCH_3$	+ 0 nm
O-alkyl	+ 6 nm
S-alkyl	+30 nm
Cl, Br	+ 5 nm
$N(alkyl)_2$	+60 nm

solvent correction   ∿ 0 nm

Example:

Estimating the absorption maximum for

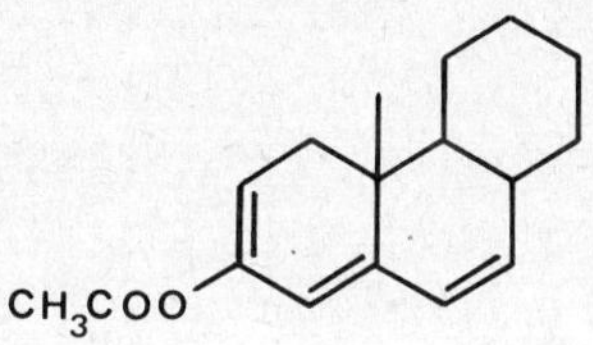

CH_3COO—

	nm
base value (homoannular)	253
additional conjugated double bond	30
exocyclic double bond	5
3 C-substituents	15
estimated:	303 nm
determined:	306 nm

UV/VIS

AROMATIC CARBONYL COMPOUNDS

<u>UV-Absorption of Aromatic Carbonyl Compounds</u>

{Scott rules to estimate the position of the K-band (in ethanol as the solvent)}

Parent system:

⬡–CO-R   R : alkyl, alicyclyl   246 nm

⬡–CO-H   250 nm

⬡–CO-OH   230 nm

⬡–CO-OR   230 nm

Increments (in nm) per substituent in	ortho	meta	para
alkyl, alicyclyl	3	3	10
OH, O-alkyl	7	7	25
O^-	11	20	78
Cl	0	0	10
Br	2	2	15
NH_2	13	13	58
$NHCOCH_3$	20	20	45
$N(CH_3)_2$	20	20	85

<u>Example:</u>

Estimating the position of the absorption maximum (K-band) for

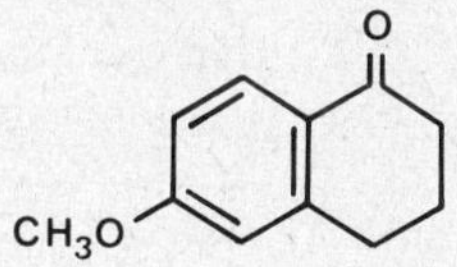

	nm
base value:	246
alicyclyl in o-position	3
O-alkyl in p-position	25
	———
estimated:	274 nm
determined:	276 nm

UV-Absorption of Aromatic Compounds

$\langle\!\!\rangle$—R Substituent R (solvent)	Transitions							
	$\pi \to \pi^*$ (allowed)		$\pi \to \pi^*$ ("forbidden")		$\pi \to \pi^*$ (substituent delocalized by aromatic system); K-band		$n \to \pi^*$ (substituent with non-bonding electron pair); R-band	
	$\lambda_{max}[nm]$	ε	$\lambda_{max}[nm]$	ε	$\lambda_{max}[nm]$	ε	$\lambda_{max}[nm]$	ε
	$\sim$180–230	$\sim$2000–10000	$\sim$250–290	$\sim$100–2000	$\sim$220–250	$\sim$10000–30000	$\sim$275–350	$\sim$10–100
-H (cyclohexane)	198	8000	255	230				
-CH$_3$ (hexane)	208	7900	262	230				
-OH (water)	211	6200	270	1450				
-O$^-$ (water)	235	9400	287	2600				
-NH$_2$ (water)	230	8600	280	1430				
-NH$_3^+$ (water)	203	7500	254	160				
-NO$_2$ (hexane)	208 / 213	9800 / 8100	270	800	251	9000	322	150
-Cl (ethanol)	210	7500	257	170				
-CH=CH$_2$ (ethanol)			282	450	244	12000		
-C≡CH (hexane)			278	650	236	12500		
-COCH$_3$ (ethanol)			278	1100	243	13000	319	50
-CHO (hexane)			280	1400	242	14000	$\sim$330	$\sim$60
-COOH (water)	202	8000	270	800	230	10000		
-C≡N (water)			271	1000	22	13000		

UV / VIS

$$CH_3-CH=CH-CH=CH_2$$

(in heptane)

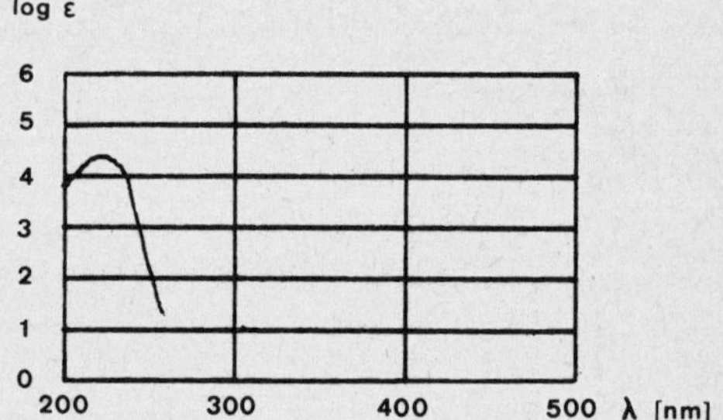

$$CH_3,\ H$$
$$C=C$$
$$H,\ COCH_3$$

(in ethanol)

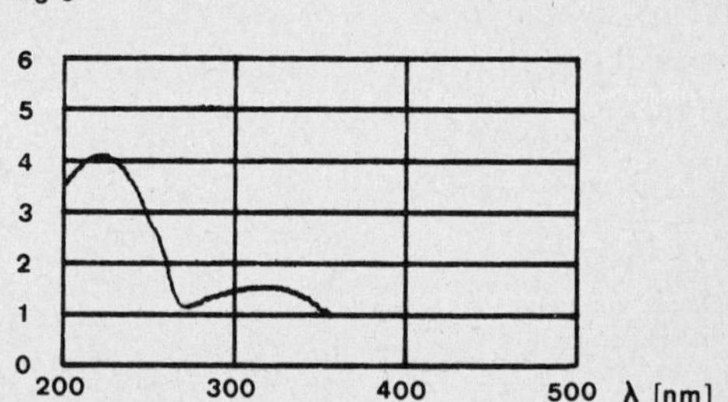

$$CH_3,\ H$$
$$C=C$$
$$H,\ COOH$$

(in water)

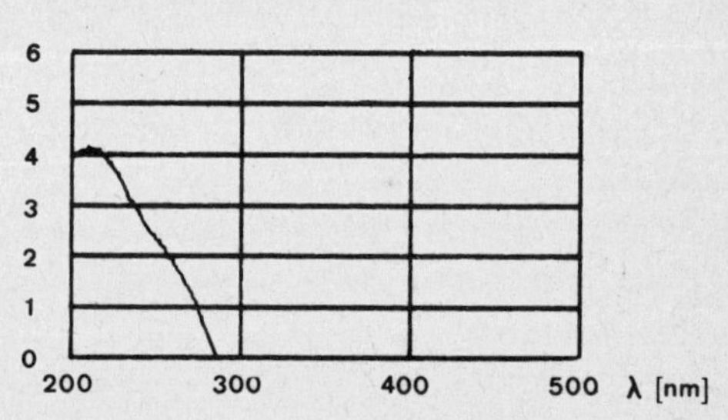

$$CH_3-NO_2$$

(in hexane)

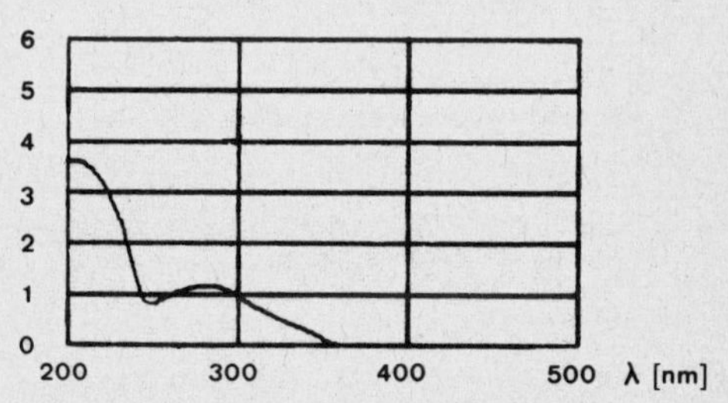

$CH_3CH_2COCH_3$

(in heptane)

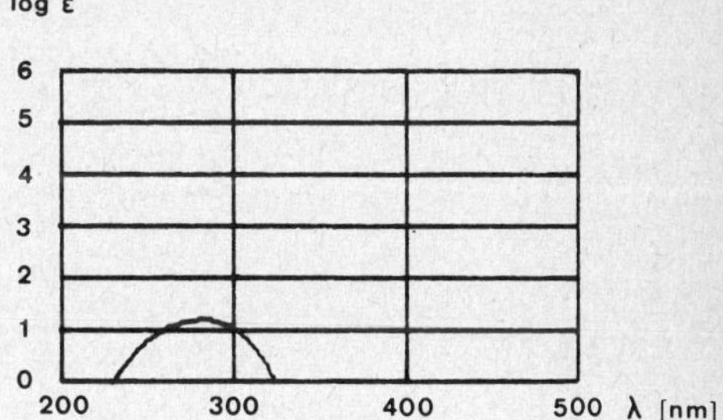

$$\begin{array}{c} CH_3 \\ \end{array} \diagdown \begin{array}{c} H \\ C=C \\ \end{array} \diagup \begin{array}{c} \\ CHO \end{array}$$

(in hexane)

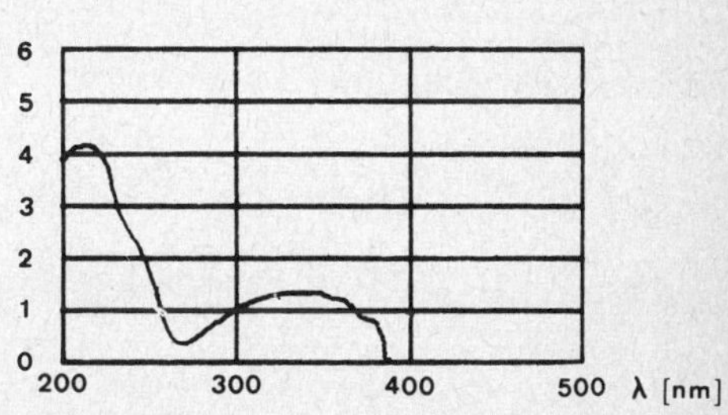

(in water)

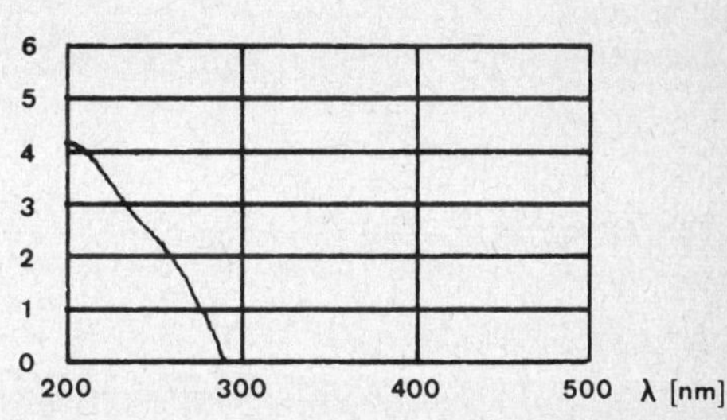

$(CH_3)_3C{-}O{-}NO$

(in hexane)

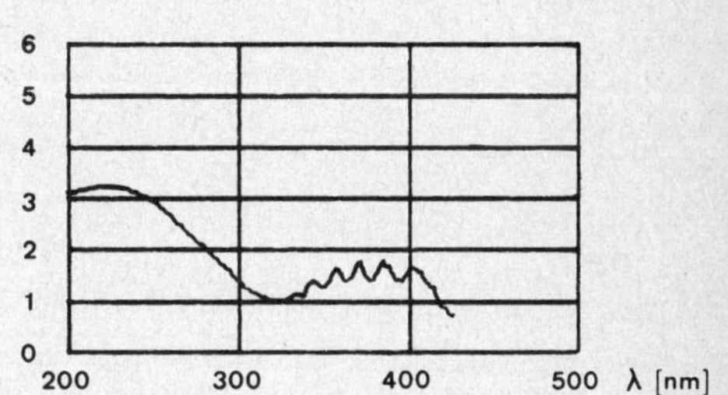

UV/ VIS

CH_3-CH_2-SH

(in heptane)

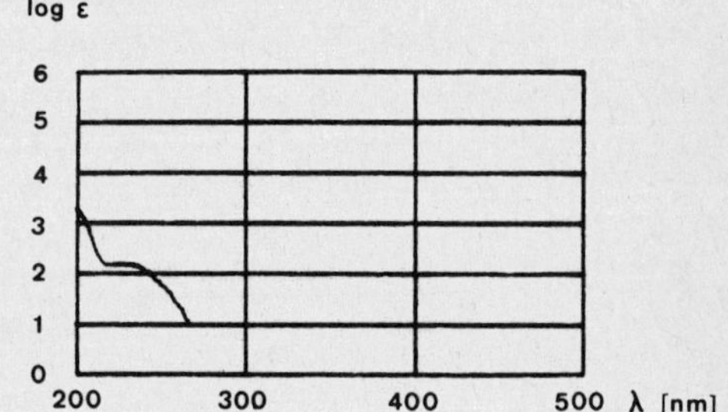

$CH_3-S-S-CH_3$

(in ethanol)

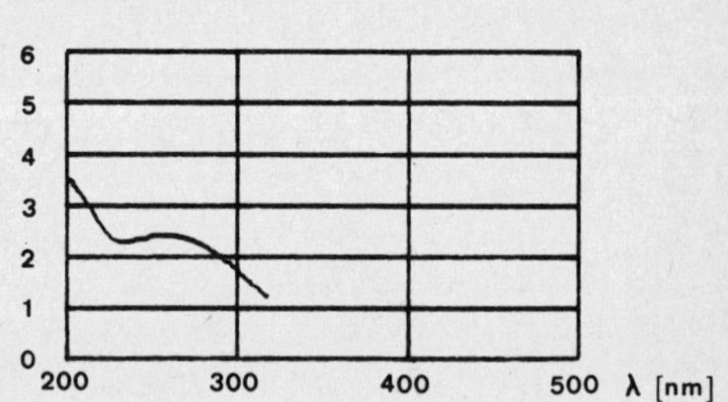

$(CH_3)_3C-Br$

(in heptane)

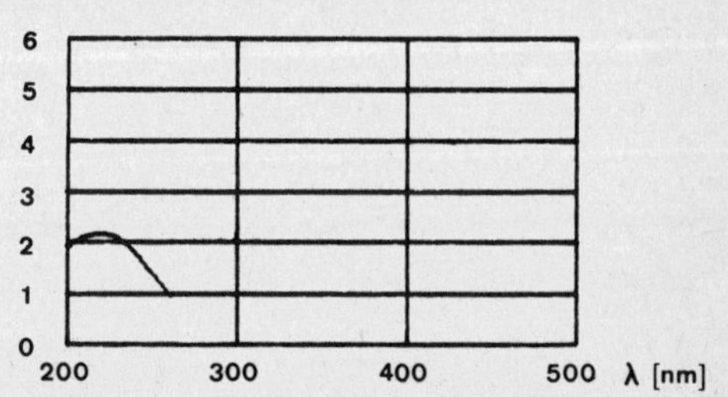

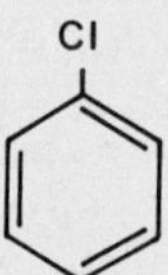

(in heptane)

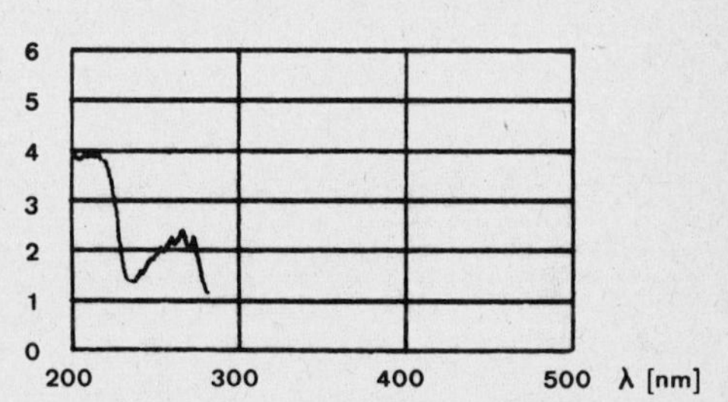

$CH_3CH_2-S-CH_2CH_3$

(in heptane)

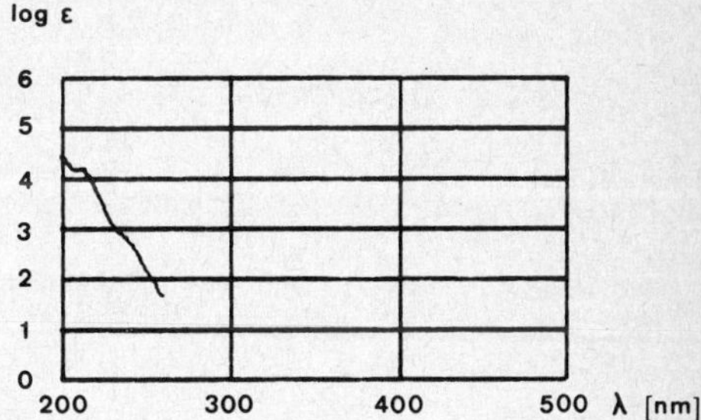

$(CH_3CH_2)_3N$

(in heptane)

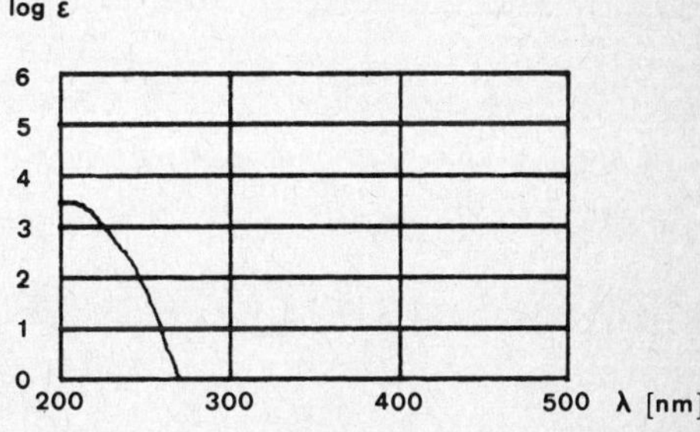

$(CH_3)_2CH-I$

(in heptane)

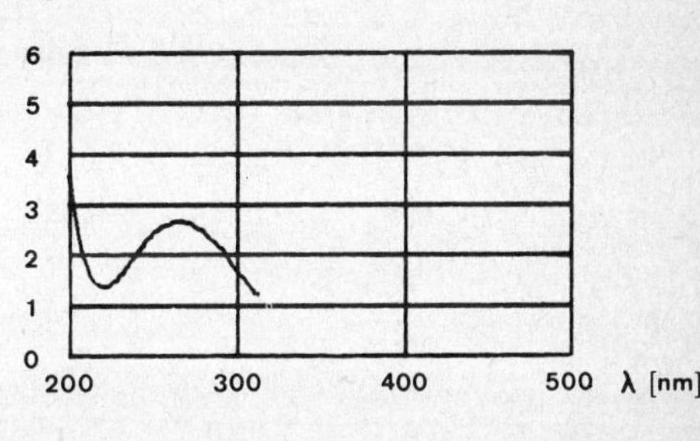

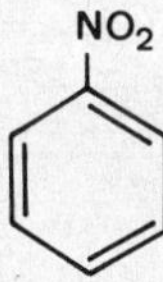

(in petroleum ether)

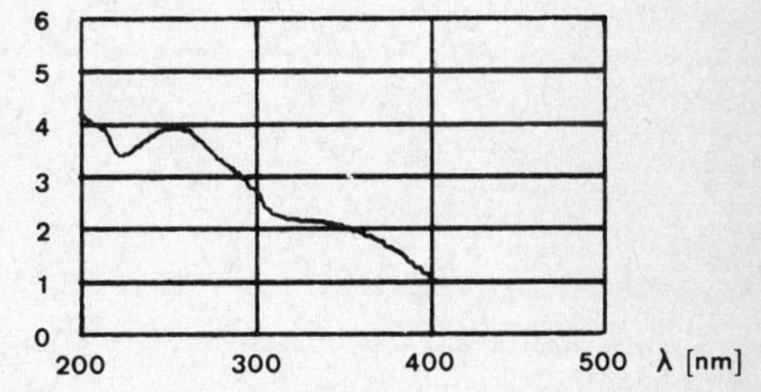

UV / VIS

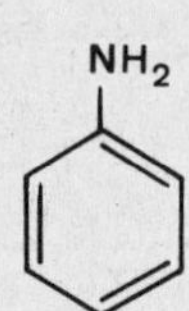

(in water)

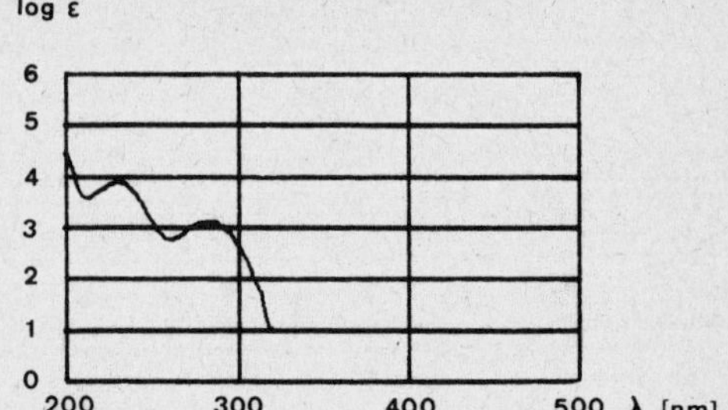

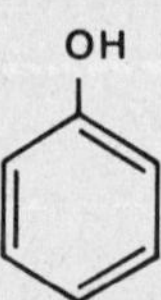

(in heptane)

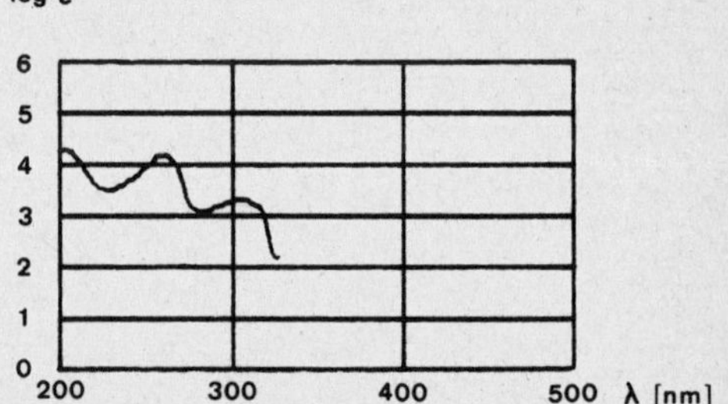

(in water)

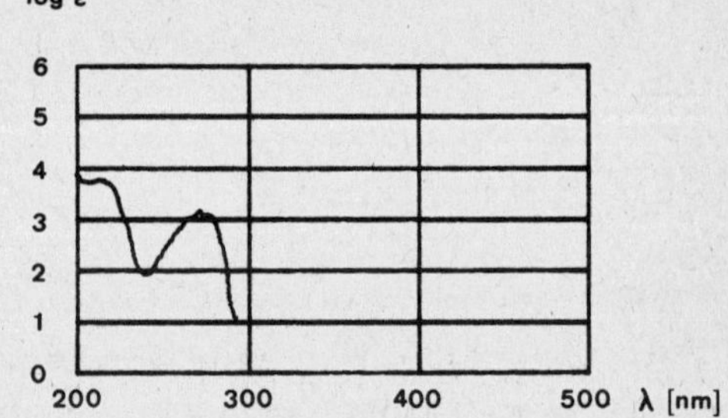

OCH₃

(in isooctane)

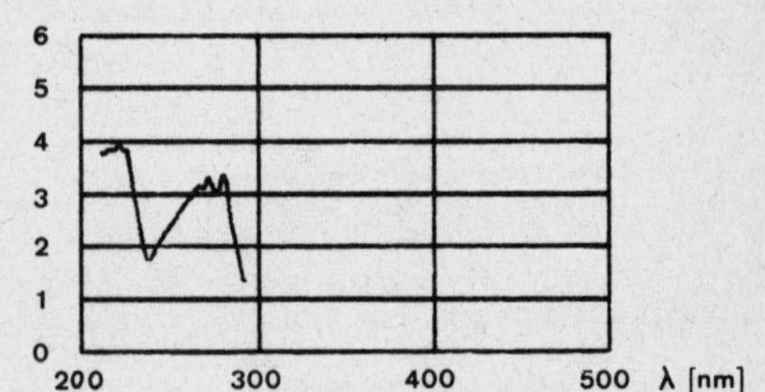

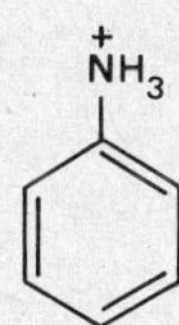

(in water)

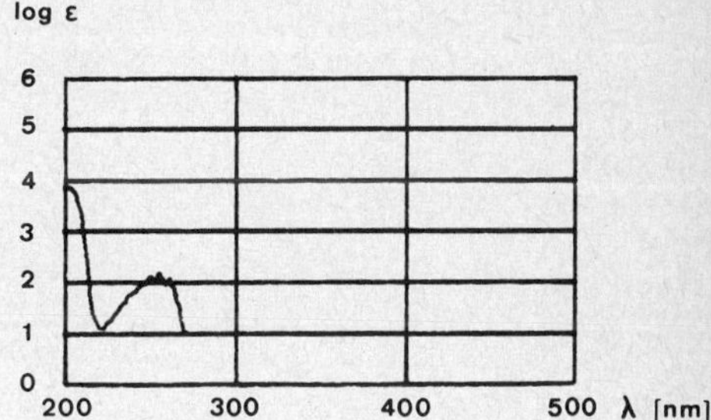

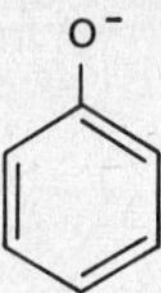

(in cyclohexane)

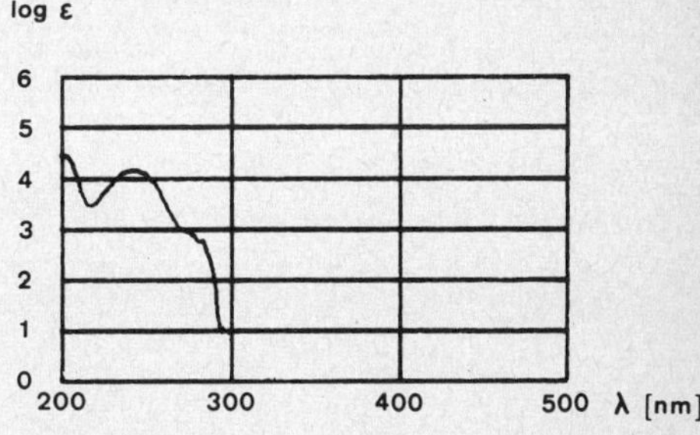

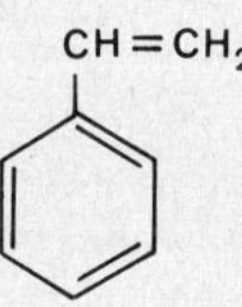

(in water)

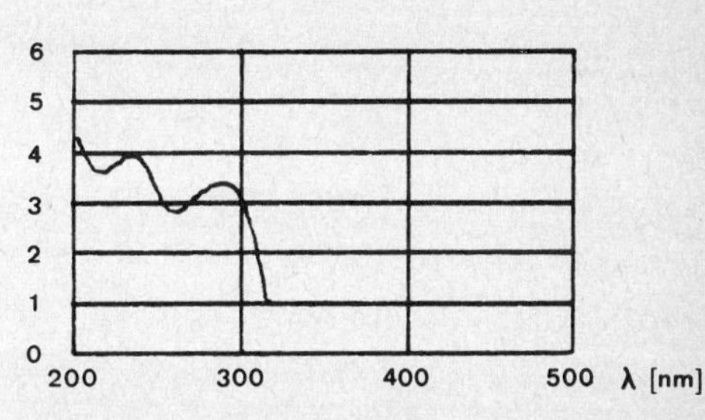

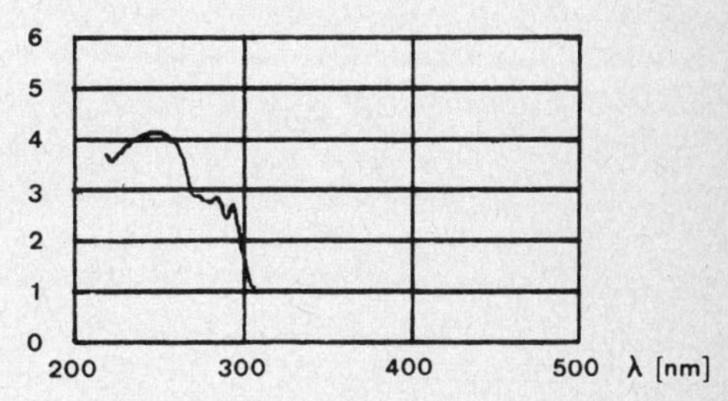

(in heptane)

UV / VIS

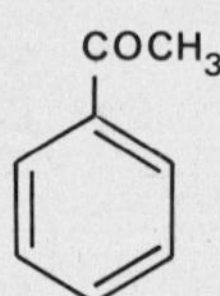

(in ethanol)

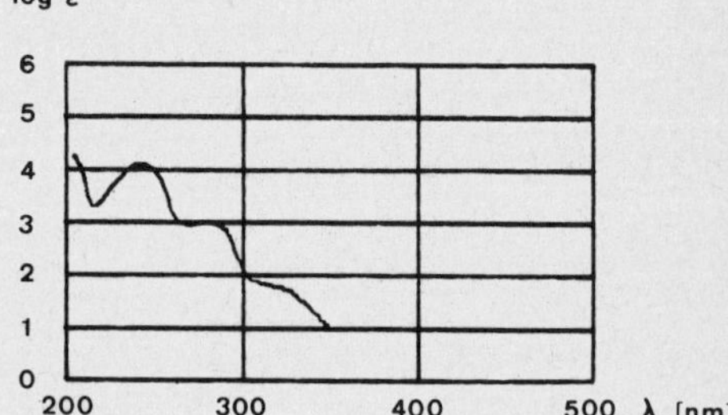

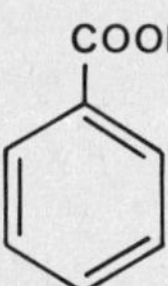

(in ethanol)

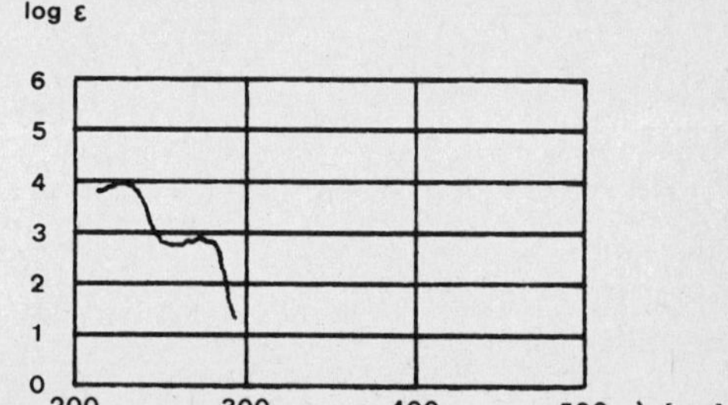

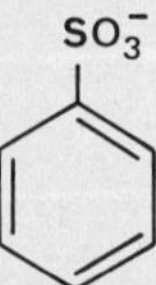

(in water)

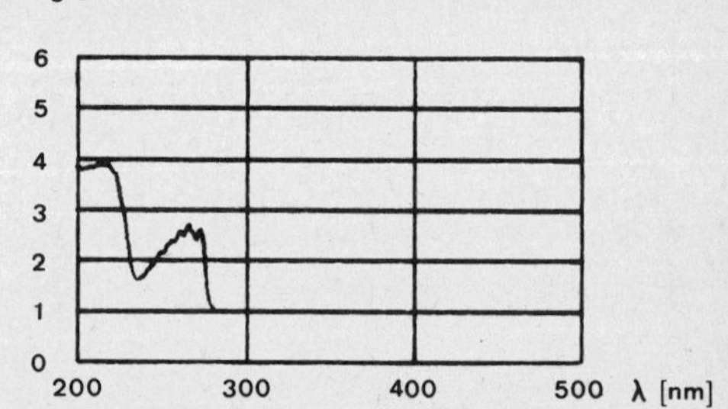

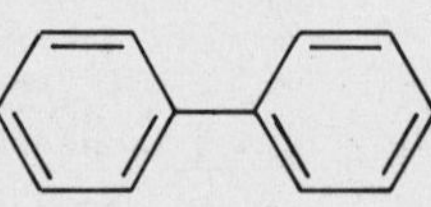

(in petroleum ether)

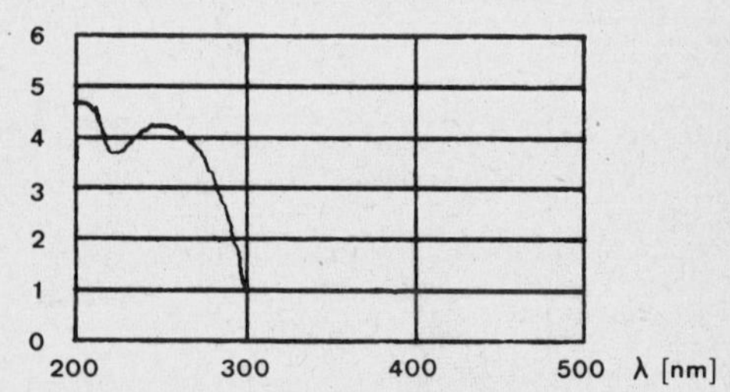

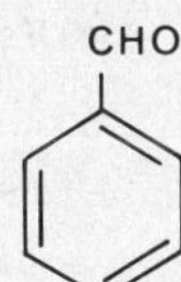

(in heptane)

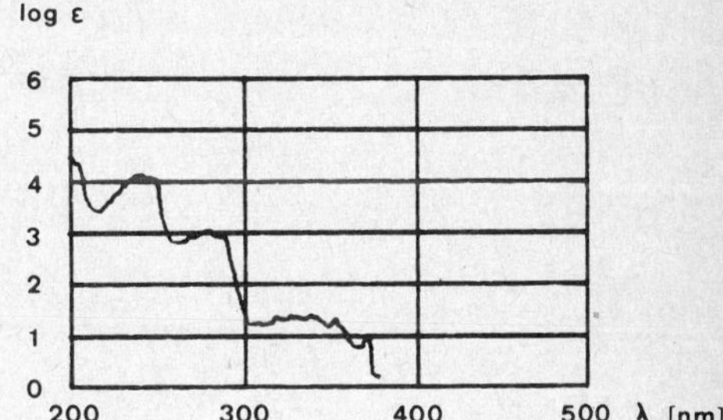

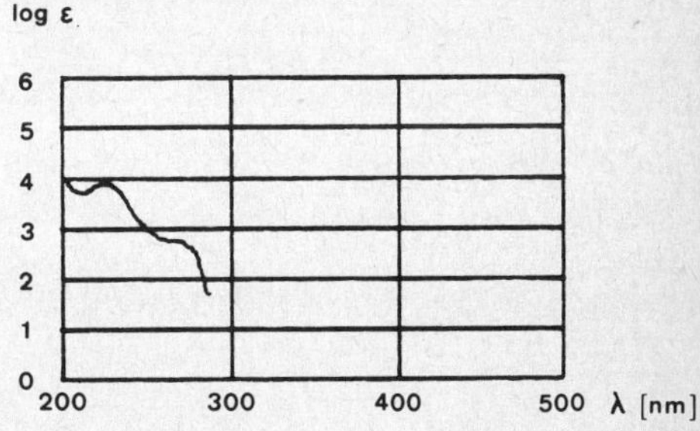

(in water)

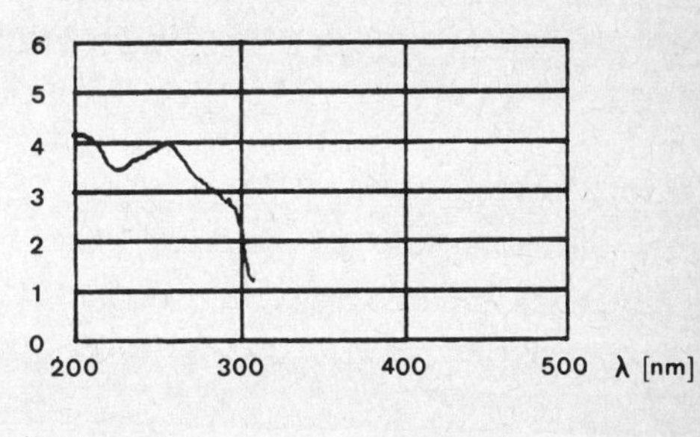

(in heptane)

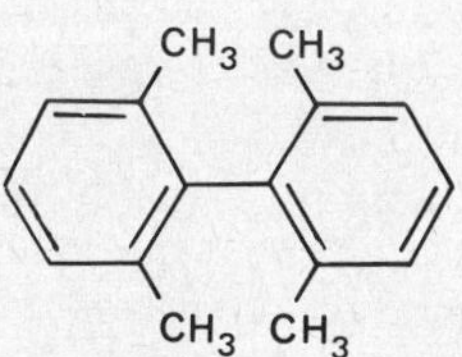

(in petroleum ether)

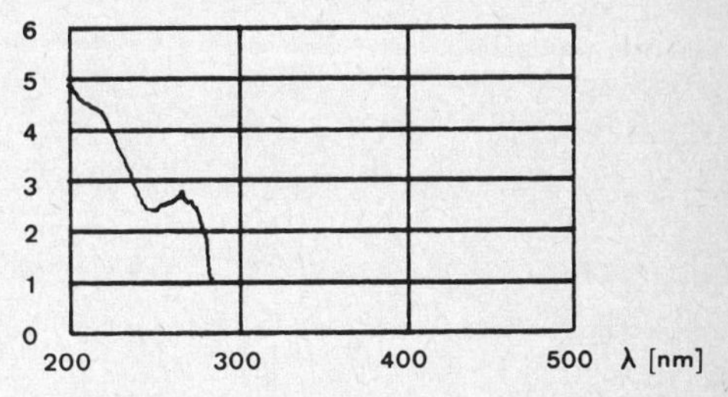

UV / VIS

REFERENCE SPECTRA

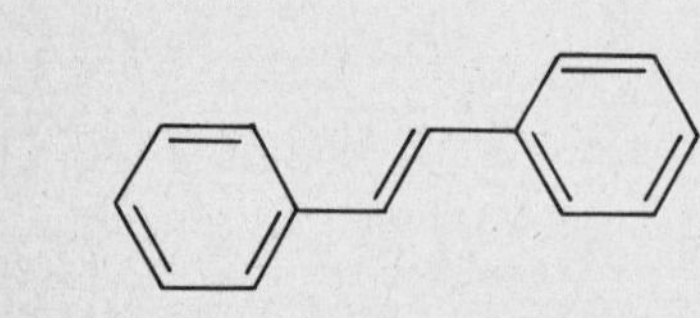

(in ethanol)

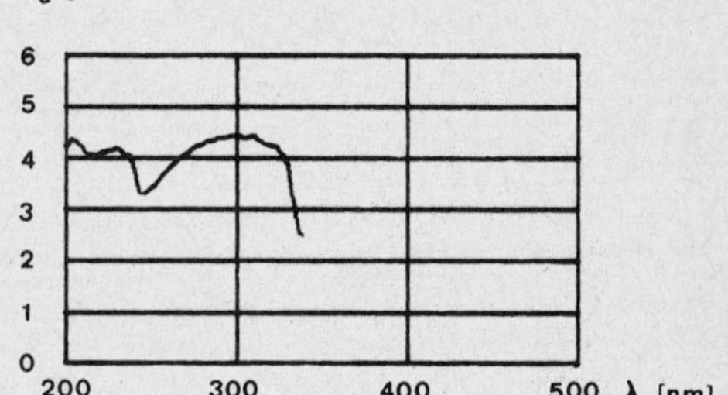

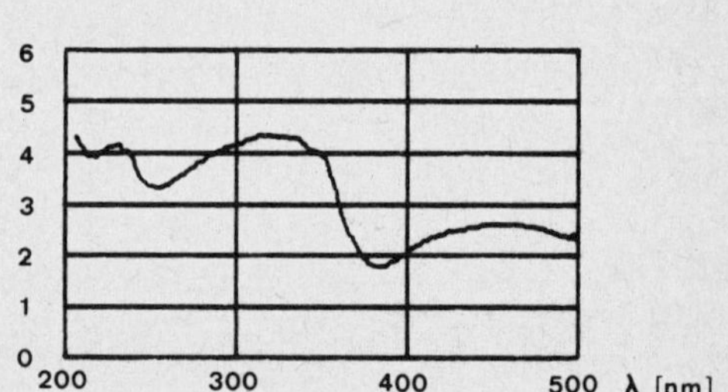

(in hexane)

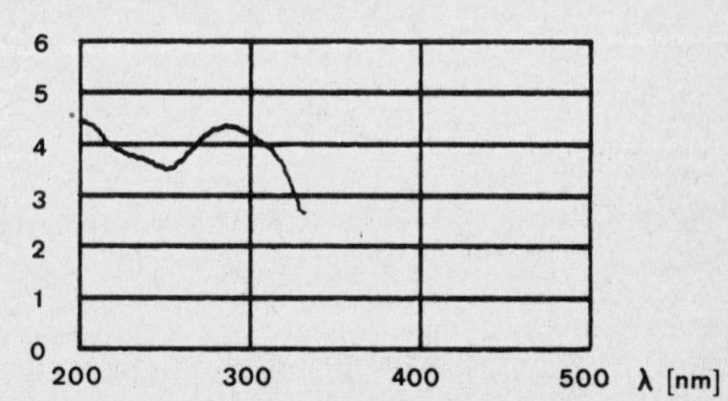

(in ethanol)

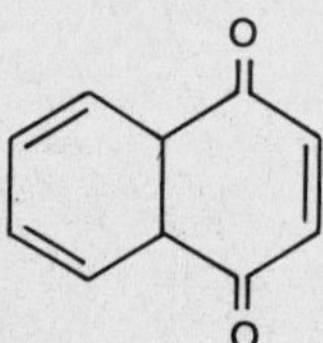

(in methanol)

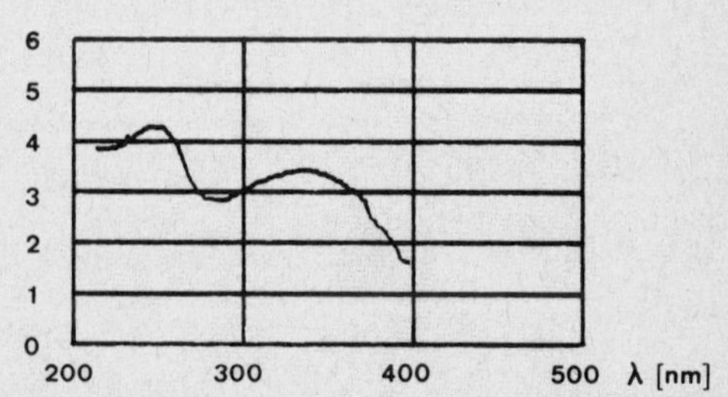

U100

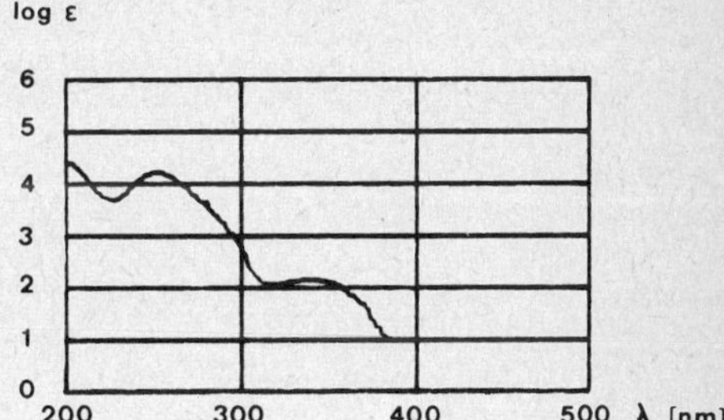

(in ethanol)

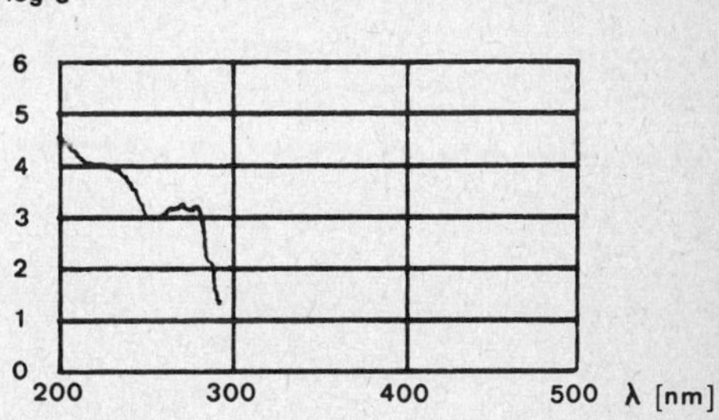

(in ethanol)

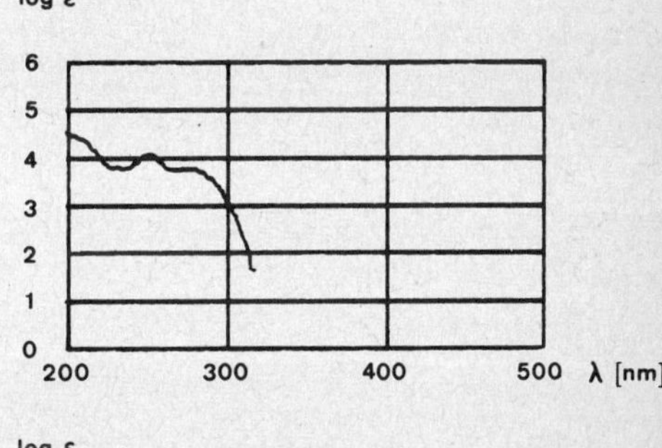

(in ethanol)

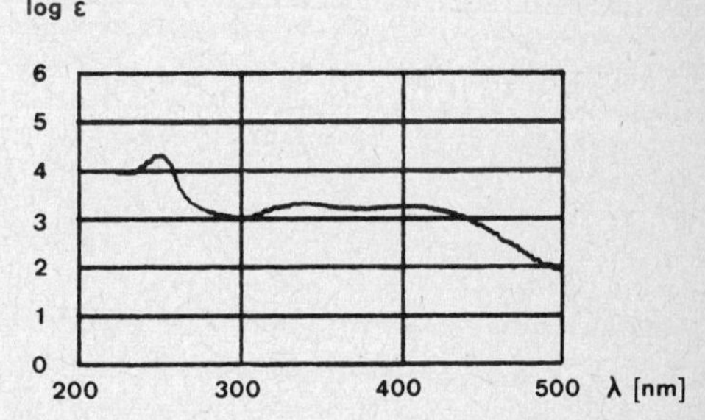

(in methanol)

UV/VIS

REFERENCE SPECTRA

U110

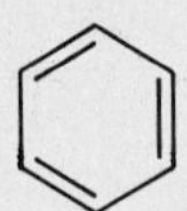

(in hexane)

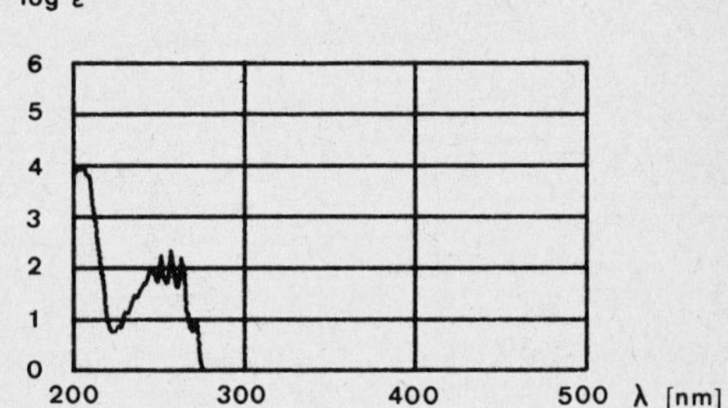

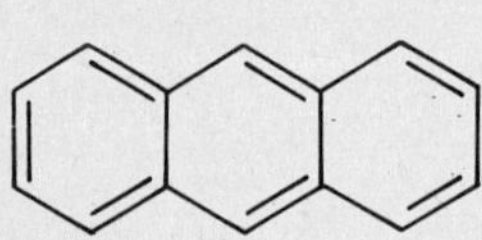

(in hexane)

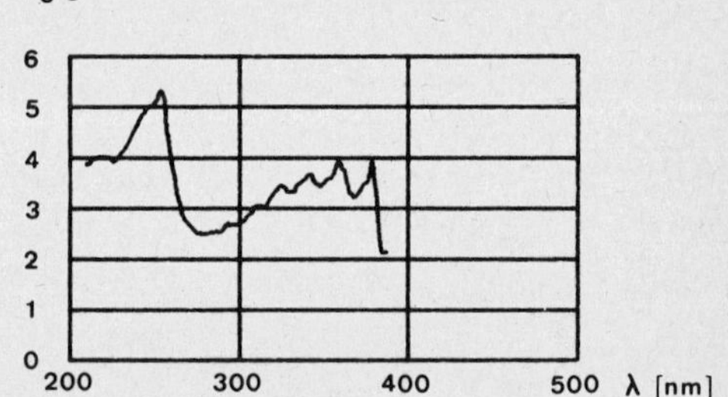

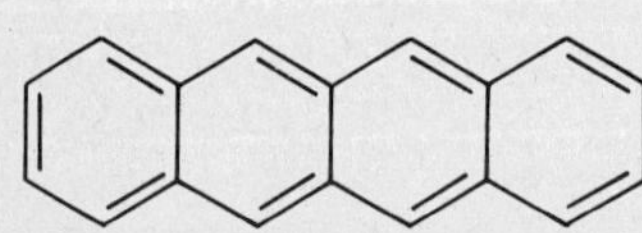

(in heptane)

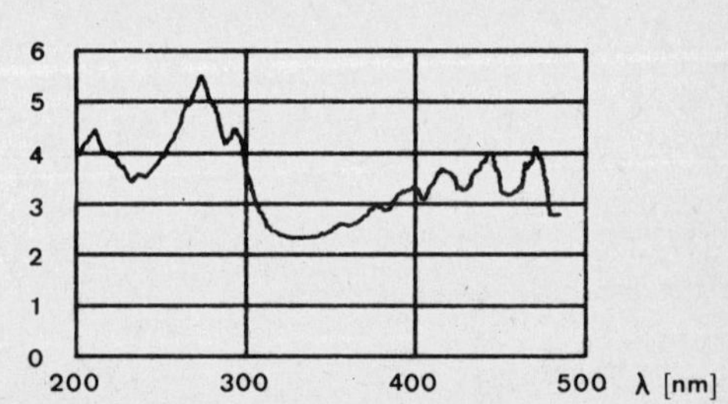

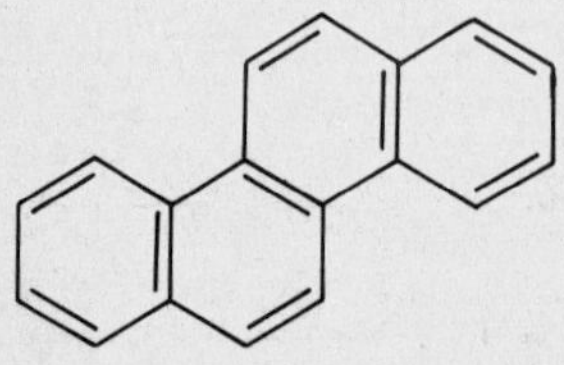

(in heptane)

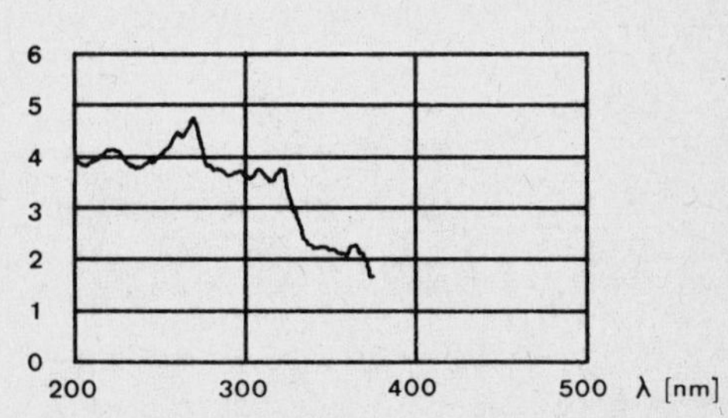

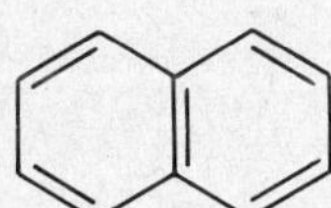

(in hexane)

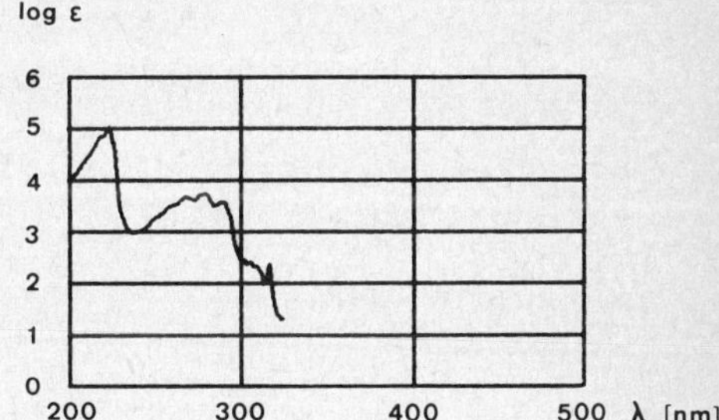

(in hexane)

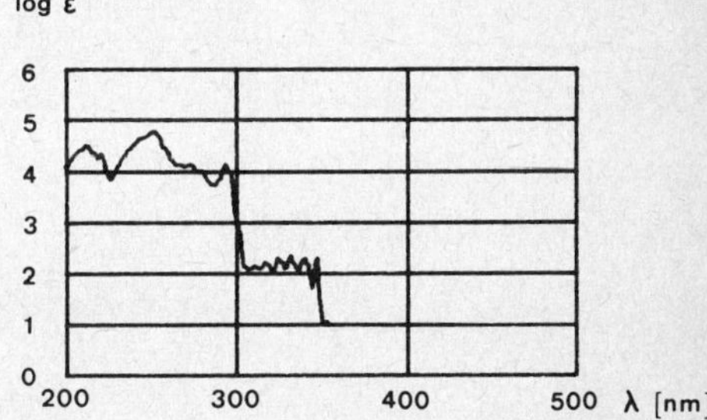

(in heptane)

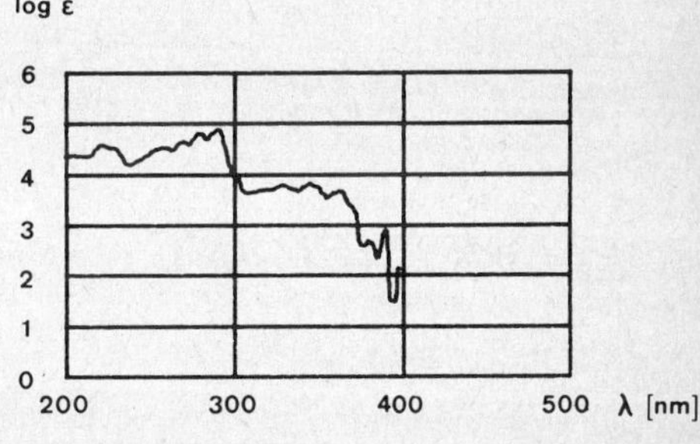

(in heptane)

UV/VIS

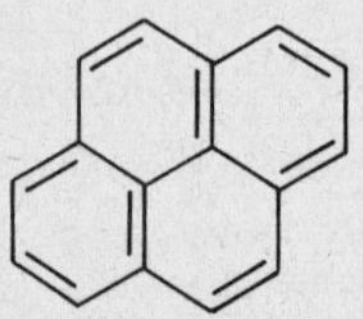

(in petroleum ether)

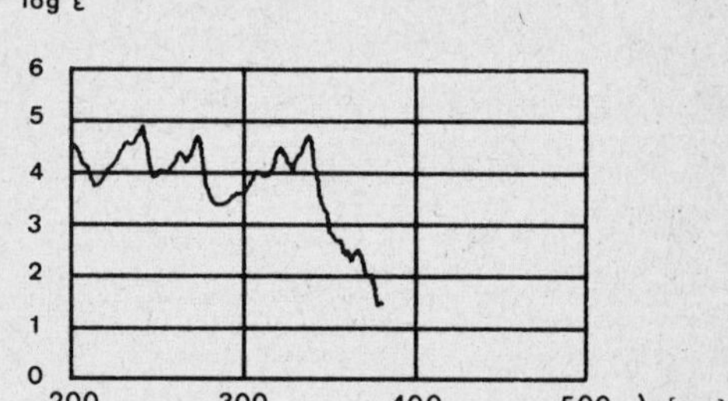

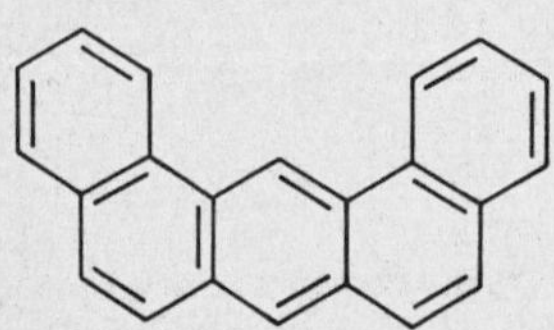

(in heptane)

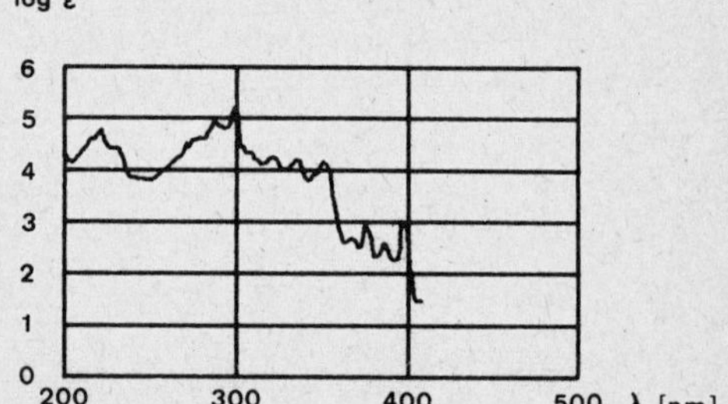

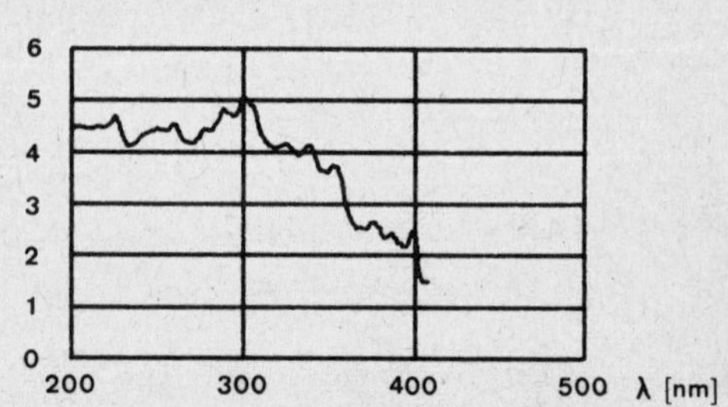

(in heptane)

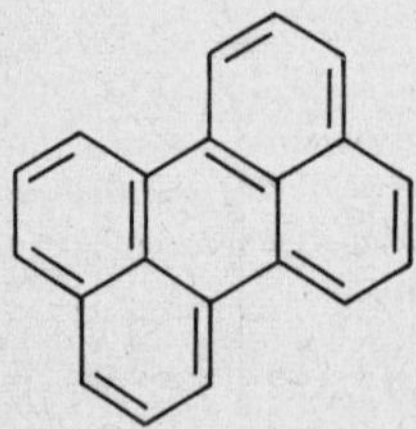

(in petroleum ether)

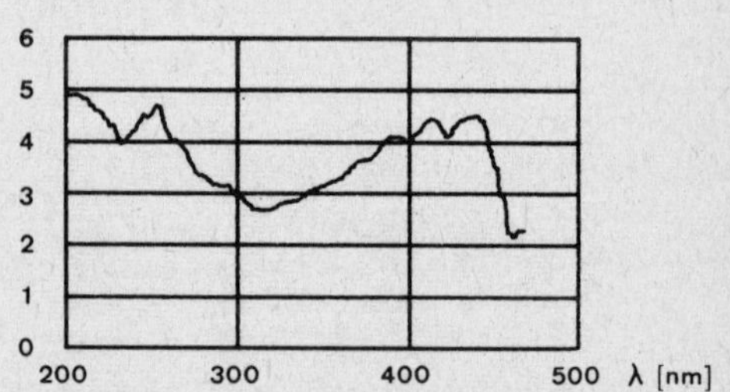

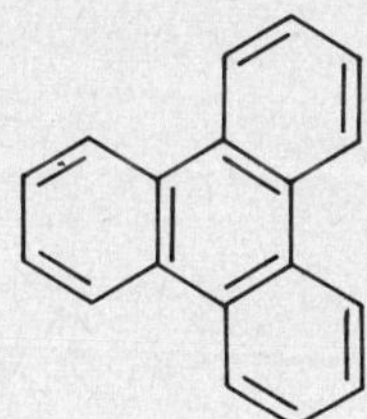

(in heptane)

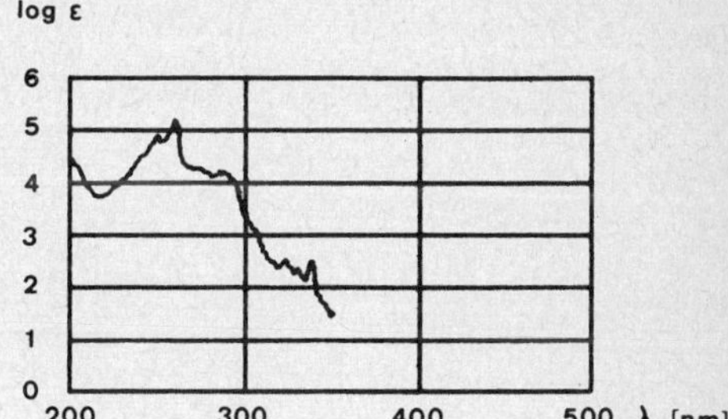

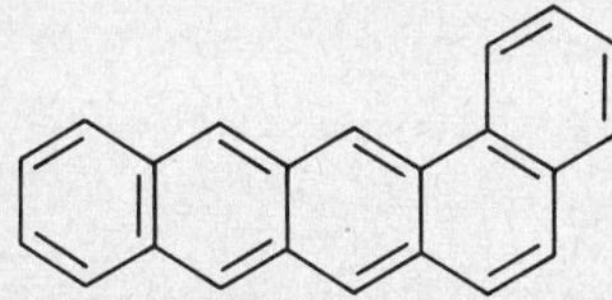

(in heptane)

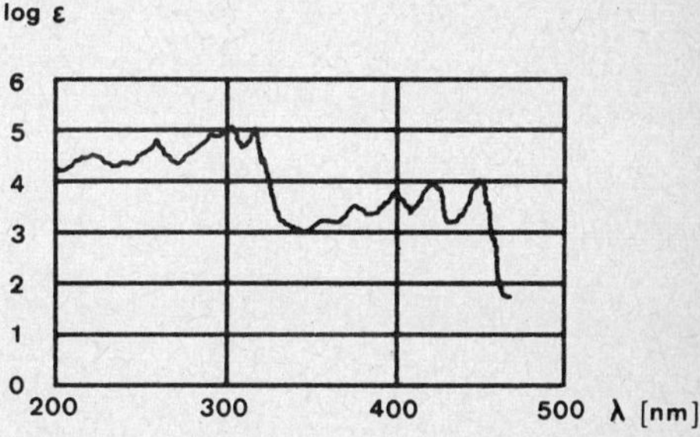

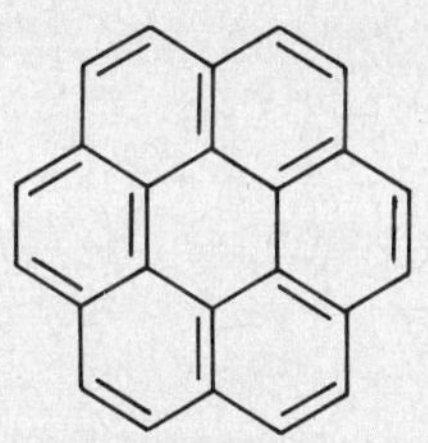

(in petroleum ether)

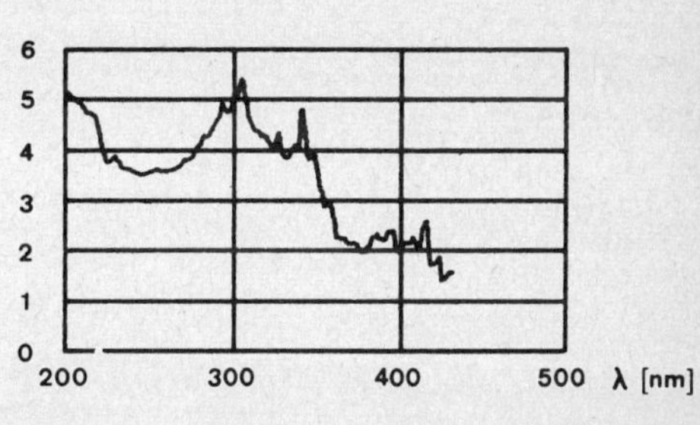

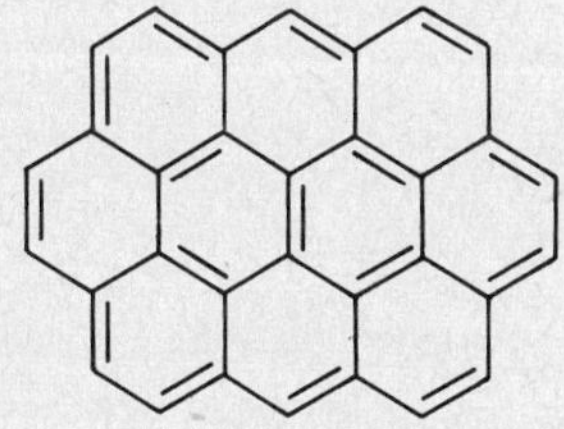

(in petroleum ether)

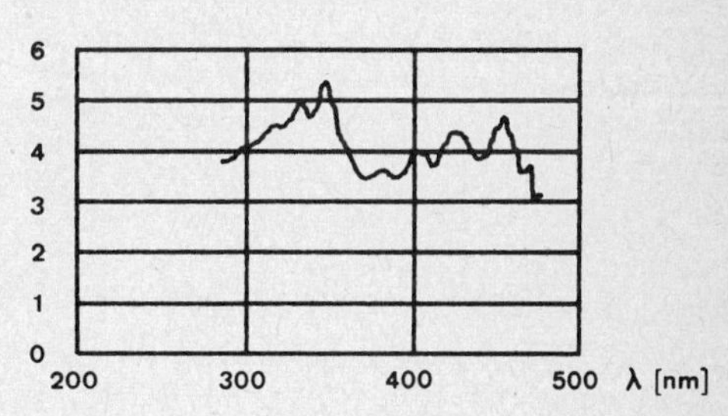

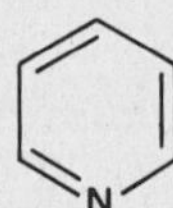

(in petroleum ether)

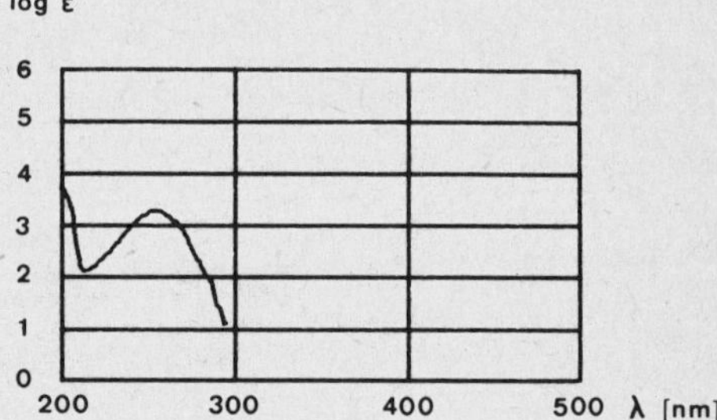

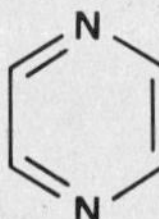

(in petroleum ether)

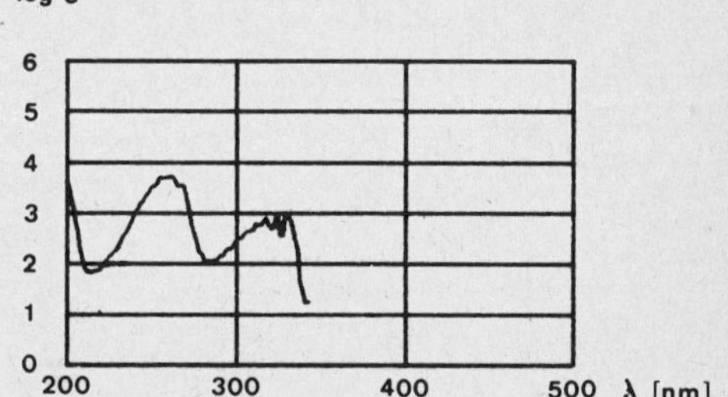

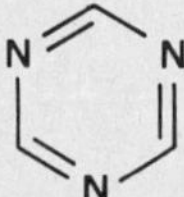

(in petroleum ether)

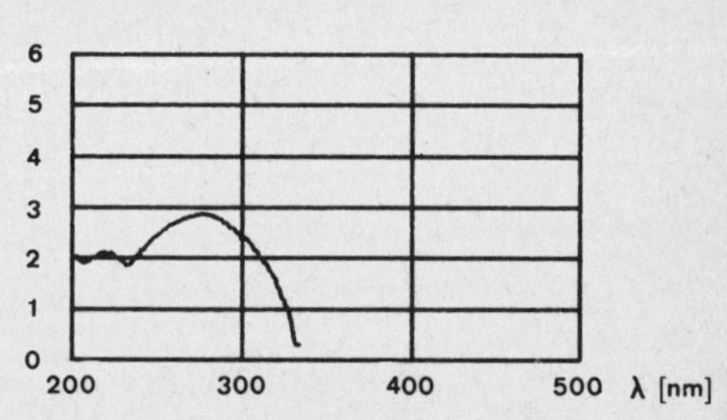

(in petroleum ether)

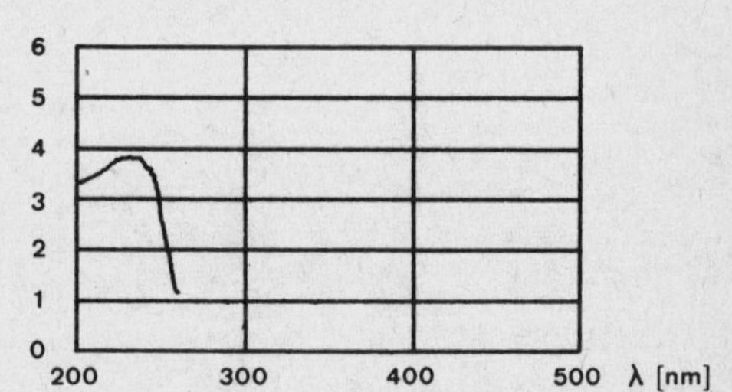

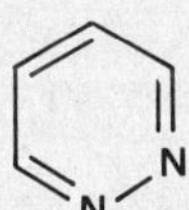

(in petroleum ether)

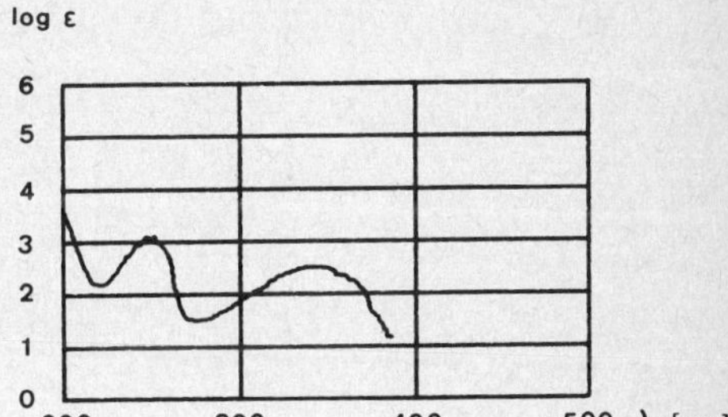

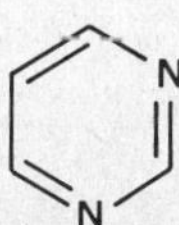

(in petroleum ether)

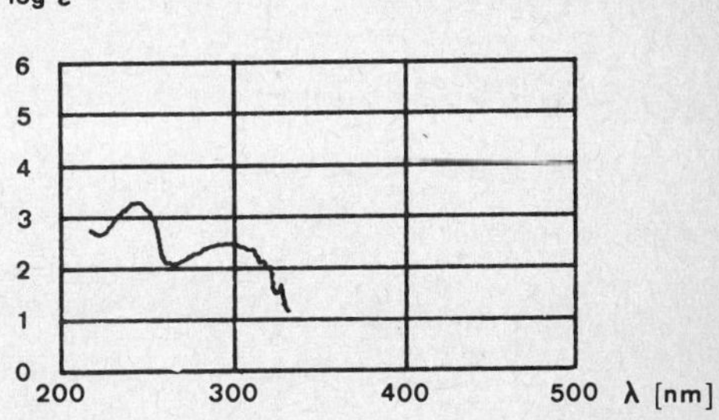

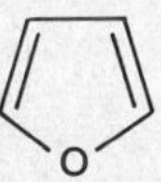

(in petroleum ether)

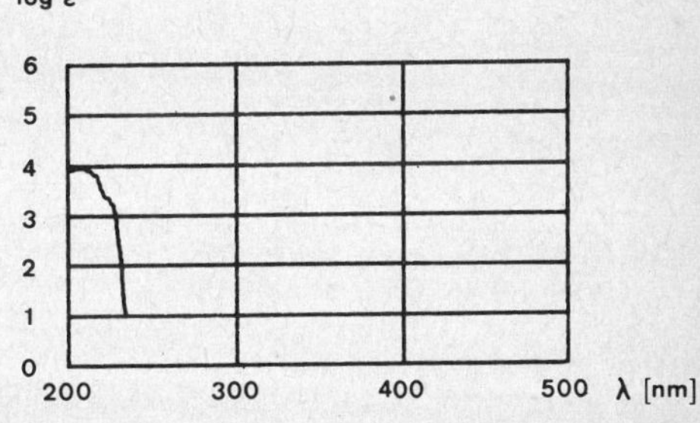

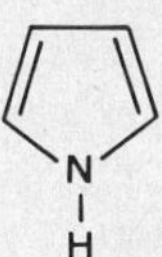

(in petroleum ether)

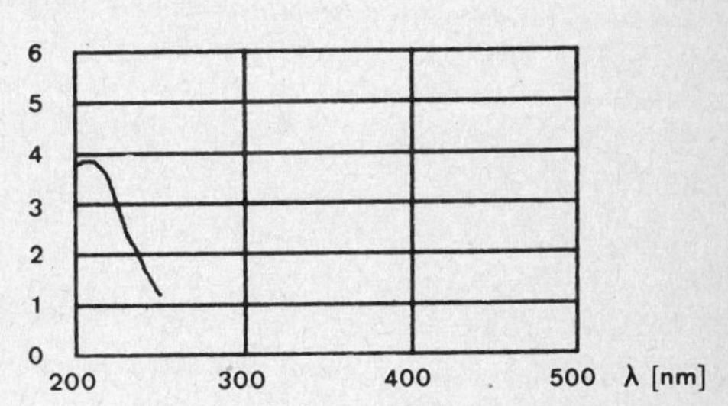

UV / VIS

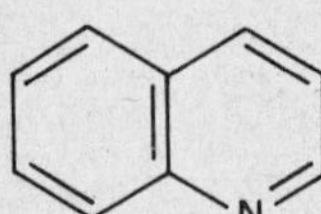

(in methanol)

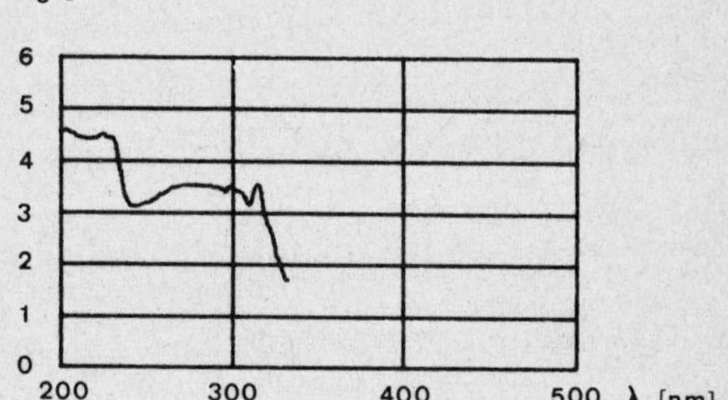

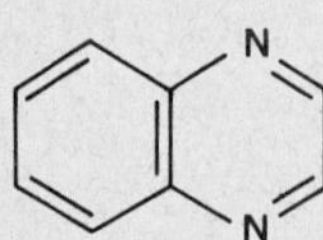

(in heptane)

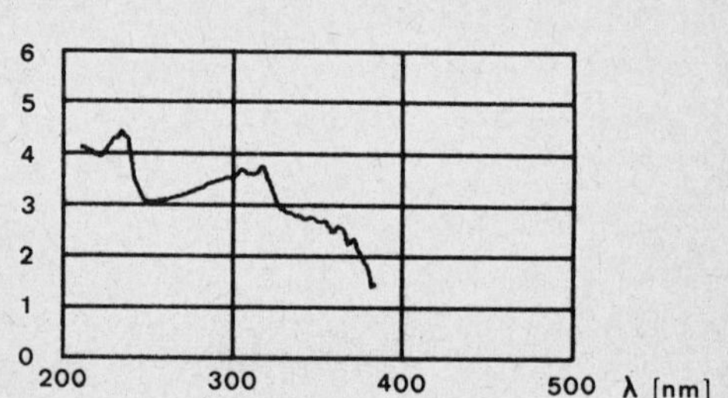

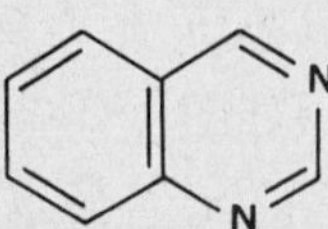

(in cyclohexane)

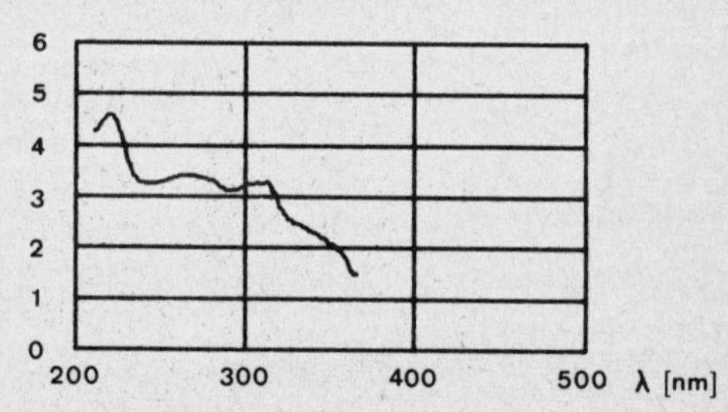

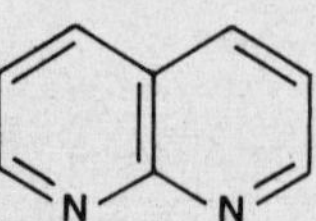

(in cyclohexane)

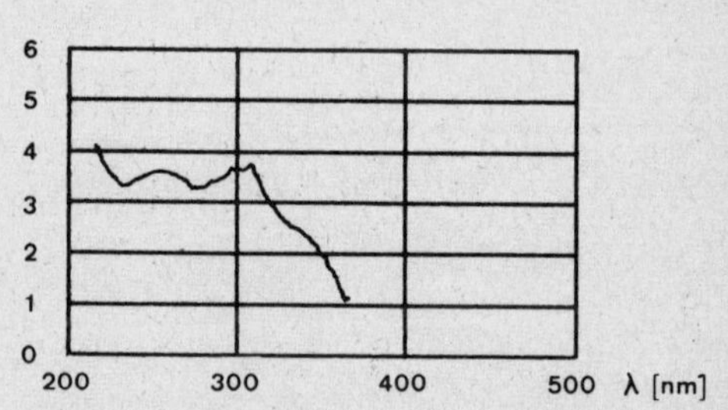

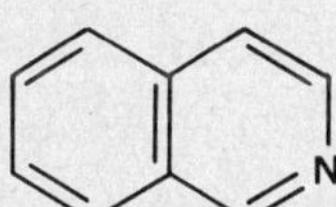

(in petroleum ether)

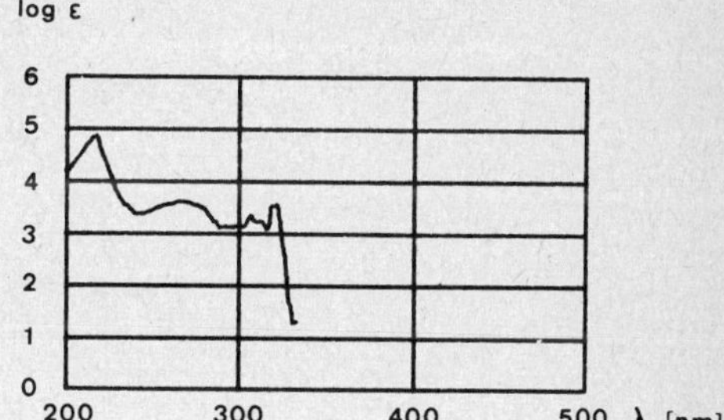

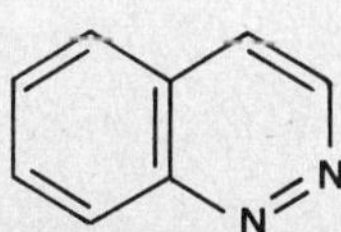

(in cyclohexane)

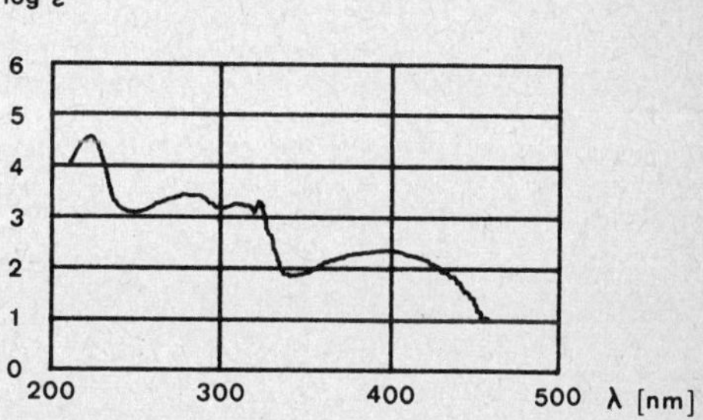

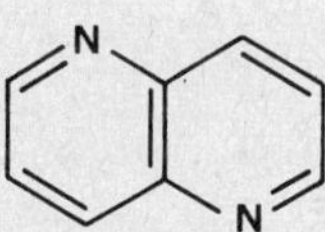

(in cyclohexane)

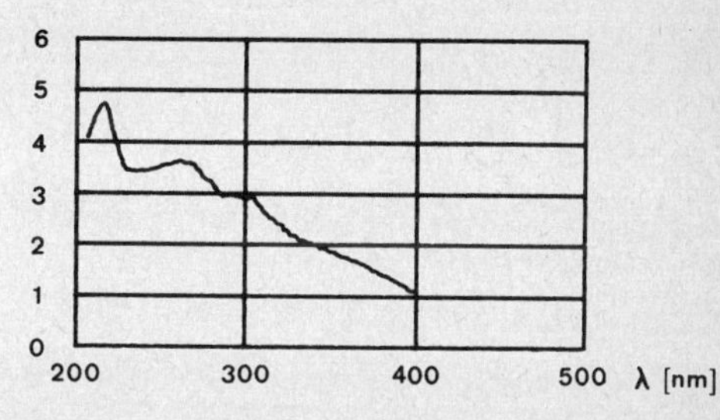

(in cyclohexane)

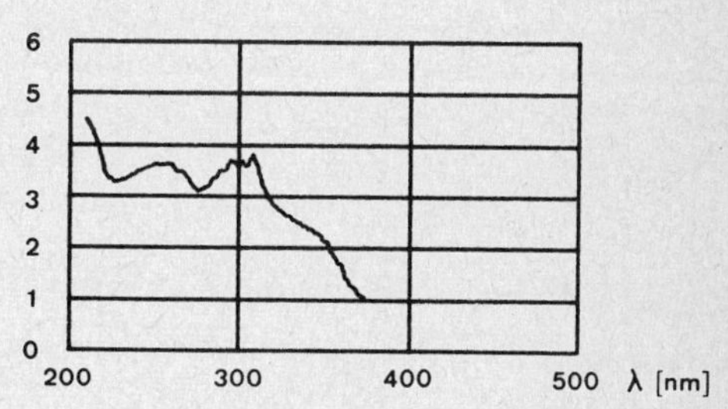

UV / VIS

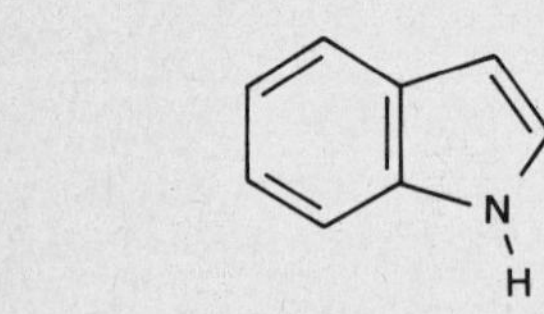

(in heptane)

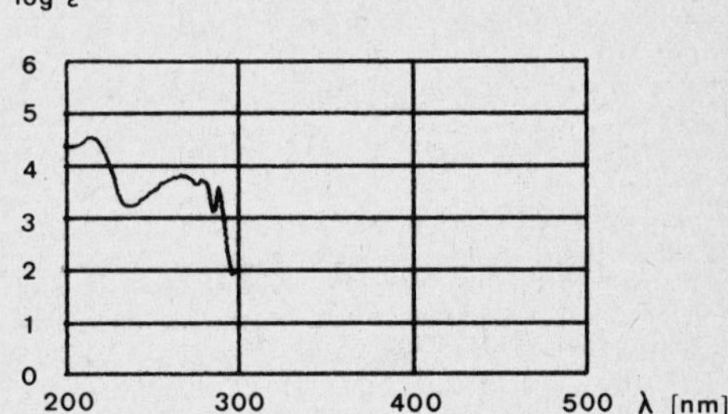

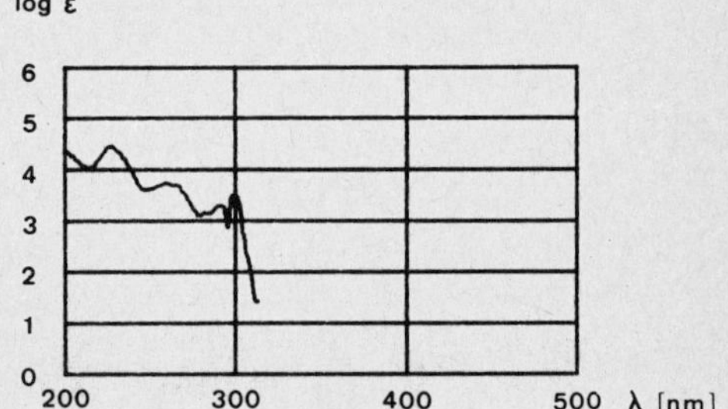

(in petroleum ether)

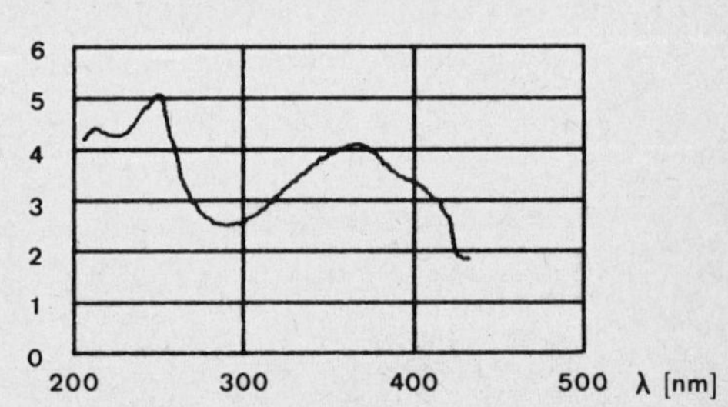

(in methanol)

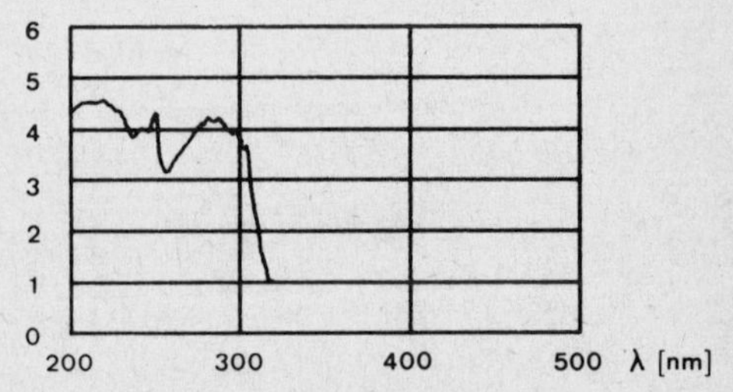

(in petroleum ether)

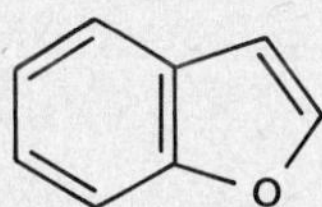

(in ethanol)

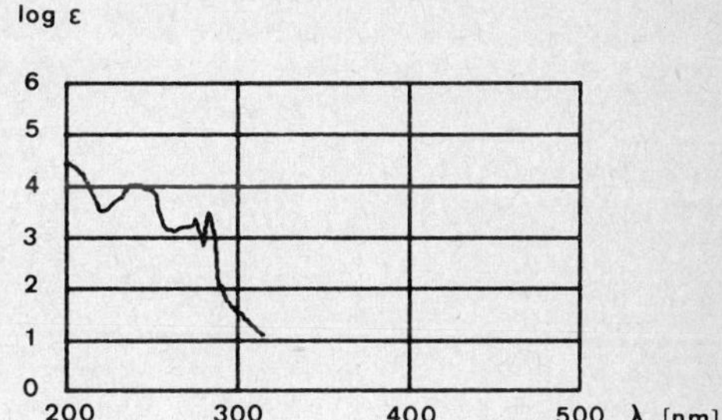

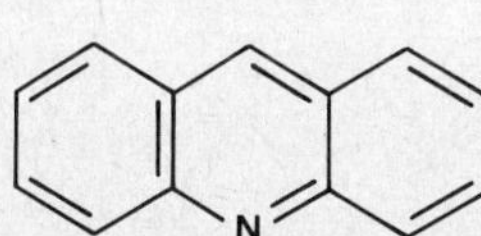

(in ethanol)

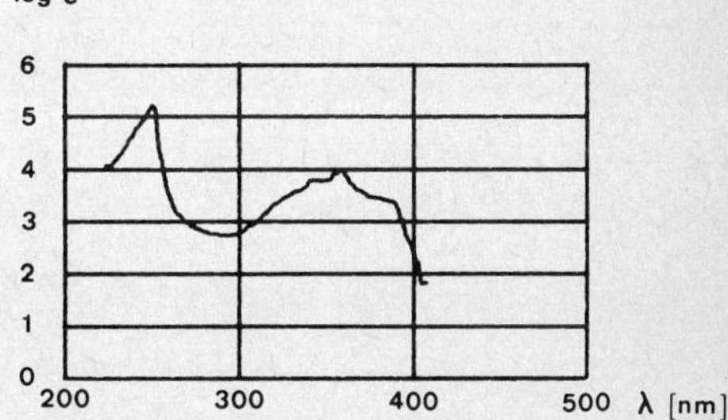

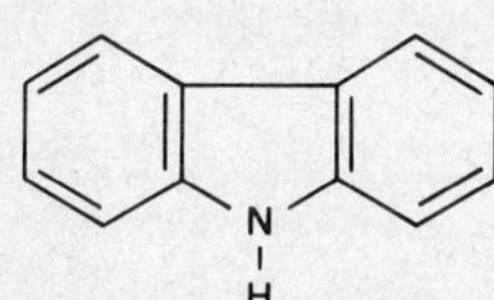

(in ethanol)

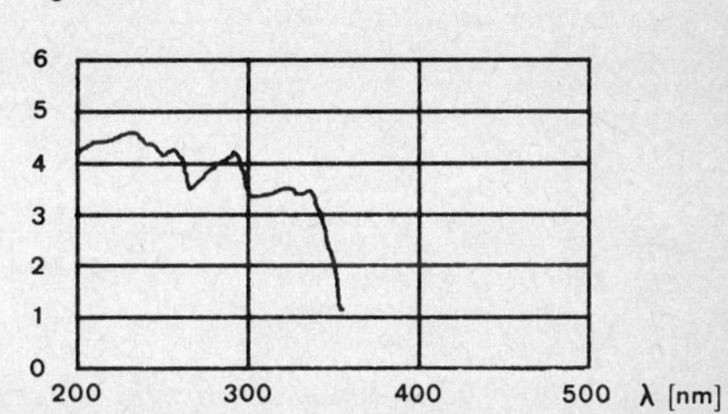

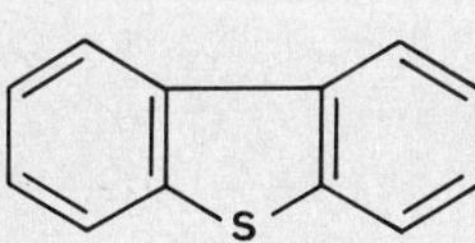

(in heptane)

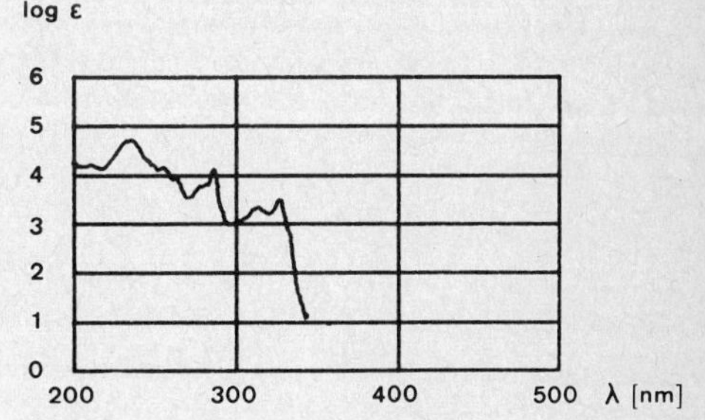

1,8-Naphthyridine U140

Neopentane C5

Nitramines I210, M125

Nitrates I210, M125, U15

Nitriles C190, I65, M10, M125
    alicyclic C190, I65
    aliphatic C10, C30, C35,
        C190, C250, H10, H15, H180
        H365, I10, I65, M155, U15
    alkynyl- I140
    aromatic C125, C140, C190,
        H260, H315, I65, U50
    olefinic C90, C190, H220,
        I30, I65

Nitriloxides I65, M125

Nitrites C10, C35, I205, M125,
    U15, U65

Nitrobenzene C125, H260, U50, U75

Nitromethane C30, H90, U60

C-Nitroso compounds I205, M10,
    M125, U15
    aromatic C125, H260, I205
    dimers I200

N-Nitroso compounds H90, I205,
    M10, M125

Nitro compounds I210, M5, M10,
    M15, M20, M125
    alicyclic C70, H195, I210
    aliphatic C10, C30, C35,
        H5, H15, H90, I10, I210,
        U15, U60
    aromatic C125, C140, H260,
        H315, I210, U50, U75
    combination table B165
    olefinic C90, H220, I210

Norbornadiene C100

Norbornane C75, H190

Norbornene C100, I35

Nujol I280

Octanes C30, C35, C170

Olefins (s. Alkene)

Opaque regions I265, I270, I275

Ovalene U125

Oxalic acid C180

Oxane C40, H70, I90, M40

Oxazole H270
    benz- C160, H330

Oxetane C40, H65, I90

N-Oxides H275, H280, H335, H340,
    M5, M125

Oximes C10, C195, H10, H175, H315,
    I195, M125, U15

Oxiranes C10, C40, H65, I30, I35,
    I90, M5, M125

Oxolan (s. Tetrahydrofuran)

Oxolenes C40, H65

Oxopyrans (s. Pyrones)

Pentadeuteropyridine C265, H370

1,3-Pentadiene U60

Pentane C5, M160

3-Penten-2-one U60

Peracids I95, M125

Perfluoroethyl derivatives M45

Perfluoroalkyl derivatives M45

Peroxides I95, M125
    cyclic M15
    diacyl I95

Perylene U120

Phenanthrene C115, H250, U115
    benzo[c]- U115
    9,10-dihydro- H240

Phenazine C165, U150

Phenetole H60, H255

Phenol C120, H50, H255, I85, M10,
    M40, M120, M125, U50, U80, U85
    derivatives M40
    combination table B135

Phenylacetylene C110, C120,
    H225, H255, U50

Phenylethyl derivatives M45

Phosphates M130
    alkyl M45, M130
    ethyl M10, M150

Phosphorous compounds C245, H220,
    H260, I235, M130, M150

Phthalates M50, M170

Phthalazine C165, H345, U145

Phthalimide
    N-methyl- H165

Phthalic acid anhydride C180

Piperazine C45
    N-methyl- H85

Piperidine C45, H85
    N-alkyl- M40
    N-methyl- C45, H75